Internet of Things with Raspberry Pi and Arduino

Internet of Things with Raspberry Pi and Arduino

Rajesh Singh
Professor, School of Electronics & Electrical Engineering
Lovely Professional University, India

Anita Gehlot
Associate Professor, School of Electronics & Electrical Engineering
Lovely Professional University, India

Lovi Raj Gupta
Executive Dean
Lovely Professional University, India

Bhupendra Singh
Managing Director
Schematics Microelectronics, India

Mahendra Swain
Assistant Professor
MIET, Jammu, India

CRC Press is an imprint of the
Taylor & Francis Group, an **informa** business

CRC Press
Taylor & Francis Group
6000 Broken Sound Parkway NW, Suite 300
Boca Raton, FL 33487-2742

First issued in paperback 2021

ISBN-13: 978-0-367-24821-5 (hbk)
ISBN-13: 978-1-03-208598-2 (pbk)

Library of Congress Cataloging-in-Publication Data

Names: Gehlot, Anita, author.
Title: Internet of things with Raspberry Pi and Arduino / by Dr. Anita Gehlot, Associate Professor, School of Electronics & Electrical Engineering, Lovely Professional University, India, Dr. Rajesh Singh, Professor, School of Electronics & Electrical Engineering, Lovely Professional University, India, Dr. Lovi Raj Gupta, Executive Dean, Lovely Professional University, India, Bhupendra Singh, Managing Director, Schematics Microelectronics, India, Mahendra Swain, Assistant Professor, MIET, Jammu, India.
Description: First edition. | New York, N.Y. : CRC Press/Taylor & Francis Group, 2019. | Includes bibliographical references and index.
Identifiers: LCCN 2019032830 (print) | LCCN 2019032831 (ebook) | ISBN 9780367248215 (hardback) | ISBN 9780429284564 (ebook)
Subjects: LCSH: Internet of things. | Raspberry Pi (Computer) | Arduino (Programmable controller)
Classification: LCC TK5105.8857 .G44 2019 (print) | LCC TK5105.8857 (ebook) | DDC 004.67/8--dc23
LC record available at https://lccn.loc.gov/2019032830
LC ebook record available at https://lccn.loc.gov/2019032831

Visit the Taylor & Francis Web site at
http://www.taylorandfrancis.com

and the CRC Press Web site at
http://www.crcpress.com

Contents

Preface xi
Acknowledgments xiii
Authors xv

Section I Introduction

1. **Introduction to Internet of Things** 3
   1.1 Characteristics of IoT 3
   1.2 Design Principles of IoT 4
   1.3 IoT Architecture and Protocols 5
      1.3.1 IoT Architecture 5
      1.3.2 IoT Protocols 6
         1.3.2.1 OSI Model 6
         1.3.2.2 Organizational Levels 7
   1.4 Enabling Technologies for IoT 9
   1.5 IoT Levels 10
   1.6 IoT vs M2M 11

2. **Sensors** 13
   2.1 Sensors Classification 13
   2.2 Working Principle of Sensors 15
   2.3 Criteria to Choose a Sensor 17
   2.4 Generation of Sensors 17

3. **IoT Design Methodology** 19
   3.1 Design Methodology 19
   3.2 Challenges in IoT Design 20
   3.3 IoT System Management 20
   3.4 IoT Servers 21
      3.4.1 KAA 22
      3.4.2 SeeControl IoT 22
      3.4.3 Temboo 22
      3.4.4 SensorCloud 22
      3.4.5 Carriots 22
      3.4.6 Xively 22
      3.4.7 Etherios 23
      3.4.8 thethings.io 23
      3.4.9 Ayla's IoT Cloud Fabric 23
      3.4.10 Exosite 23

3.4.11 OpenRemote ... 23
3.4.12 Arrayent Connect TM ... 23
3.4.13 Arkessa ... 24
3.4.14 Oracle IoT Cloud ... 24
3.4.15 ThingWorx ... 24
3.4.16 Nimbits ... 24
3.4.17 InfoBright ... 24
3.4.18 Jasper Control Center ... 24
3.4.19 AerCloud ... 24
3.4.20 Echelon ... 25
3.4.21 ThingSpeak ... 25
3.4.22 Plotly ... 25
3.4.23 GroveStreams ... 25
3.4.24 IBM IoT ... 25
3.4.25 Microsoft Research Lab of Things ... 25
3.4.26 Blynk ... 25
3.4.27 Cayenne APP ... 26
3.4.28 Virtuino APP ... 26

Section II Basics of Arduino and Raspberry Pi

4. Basics of Arduino ... 29
4.1 Introduction to Arduino ... 29
4.1.1 Arduino Uno ... 29
4.1.2 Arduino Mega ... 30
4.1.3 Arduino Nano ... 31
4.2 Arduino IDE ... 33
4.2.1 Steps to Install Arduino IDE ... 33
4.3 Basic Commands for Arduino ... 37
4.4 LCD Commands ... 37
4.5 Serial Communication Commands ... 38
4.6 Play with LED and Arduino ... 38
4.6.1 Sketch ... 40
4.7 Play with LCD with Arduino ... 40
4.7.1 Sketch ... 43

5. Basics of Raspberry Pi ... 45
5.1 Introduction to Raspberry Pi ... 45
5.1.1 Raspberry Pi Components ... 45
5.2 Installation of NOOBS on SD Card ... 48
5.3 Installation of Raspbian on SD Card ... 49
5.4 Terminal Commands ... 51
5.5 Installation of Libraries on Raspberry Pi ... 52
5.6 Getting the Static IP Address of Raspberry Pi ... 52

5.7 Run a Program on Raspberry Pi 57
5.7.1 Editing rc.local 57
5.8 Installing the Remote Desktop Server 58
5.9 Pi Camera 59
5.9.1 Testing of the Camera 59
5.9.2 Raspberry Pi Camera as a USB Video Device 59
5.10 Face Recognition Using Raspberry Pi 60
5.11 Installation of I2C Driver on Raspberry Pi 60
5.12 Serial Peripheral Interface with Raspberry Pi 63
5.13 Programming a Raspberry Pi 63
5.14 Play with LED and Raspberry Pi 64
5.14.1 Recipe for LED Blink 64
5.14.1.1 Recipe for LED Blink Using Function 65
5.15 Reading the Digital Input 66
5.15.1 Recipe 67
5.16 Reading an Edge-Triggered Input 67
5.16.1 Reading Switch in Pull-Down Configuration 67
5.16.1.1 Recipe for Pull-Down Configuration 68
5.16.2 Reading Switch in Pull-Up Configuration 69
5.16.2.1 Recipe for Pull-Up Configuration 70
5.17 Interfacing of Relay with Raspberry Pi 70
5.17.1 Recipe 71
5.18 Interfacing of DC Motor with Raspberry Pi 72
5.18.1 Recipe 73
5.18.2 Recipe to Control One Motor Using Function 74
5.19 Interfacing of LCD with Raspberry Pi 75
5.19.1 Adafruit Library for LCD 76
5.19.2 Recipe with Adafruit Library 76
5.19.3 Recipe without Library 77
5.20 Interfacing LCD with Raspberry Pi in I2C Mode 79
5.20.1 Recipe to Interface LCD in I2C Mode 81
5.21 Interfacing of DHT11 Sensor with Raspberry Pi 83
5.21.1 Recipe to Read DHT11 Sensor 85
5.21.2 Recipe to Read DHT11 Sensor and Display Data on LCD 85
5.22 Interfacing of Ultrasonic Sensor with Raspberry Pi 86
5.22.1 Recipe to Read Ultrasonic Sensor and Display Data on LCD 88
5.23 Interfacing of Camera with Raspberry Pi 90
5.23.1 Configuring the Camera with Raspberry Pi 90
5.23.2 Capturing the Image with Pi Camera 90
5.23.3 Capturing the Video with Pi Camera 93
5.23.4 Motion Detector and Capturing the Image 95
5.23.4.1 Recipe to Capture Image on Motion Detection 96

Section III Interfacing with Raspberry Pi and Arduino

6. Raspberry Pi and Arduino 99
6.1 Install Arduino IDE on Raspberry Pi 99
6.2 Play with Digital Sensor 100
6.2.1 PIR Sensor 100
6.2.2 Circuit Diagram 101
6.2.3 Sketch 102
6.3 Play with Analog Sensor 104
6.3.1 Circuit Diagram 104
6.3.2 Sketch 105
6.4 Play with Actuators 107
6.4.1 DC Motor 107
6.4.1.1 Circuit Diagram 107
6.4.1.2 Sketch 109
6.4.2 Servo Motor 110
6.4.2.1 Circuit Diagram 111
6.4.2.2 Sketch 112

7. Python and Arduino with Pyfirmata 115
7.1 Python with Arduino 115
7.2 Controlling Arduino with Python 116
7.3 Play with LED 116
7.3.1 Recipe 117
7.4 Reading an Arduino Digital Input with Pyfirmata 117
7.4.1 Recipe to Read Pull-Down Arrangement 119
7.4.2 Recipe to Read Pull-Up Arrangement 119
7.5 Reading the Flame Sensor with Pyfirmata 120
7.5.1 Program for Reading Active "Low" Flame Sensor 121
7.6 Reading an Analog Input with Pyfirmata 121
7.6.1 Recipe 122
7.7 Reading the Temperature Sensor with Pyfirmata 123
7.7.1 Recipe 124
7.8 Line-Following Robot with Pyfirmata 124
7.8.1 Recipe 125

8. Python GUI with Tkinter and Arduino 129
8.1 Tkinter for GUI Design 129
8.2 LED Blink 129
8.2.1 Recipe for LED Blinking with Fixed Time Delay 131
8.2.1.1 Tkinter GUI for LED Blinking with Fixed Delay 131
8.2.2 Recipe for LED Blinking with Variable Delay 131
8.2.2.1 Tkinter GUI for LED Blinking with Variable Delay 133

8.3 LED Brightness Control ... 133
8.3.1 Recipe ... 133
8.3.2 Tkinter GUI for LED Brightness Control ... 134
8.4 Selection from Multiple Options ... 135
8.4.1 Recipe ... 136
8.4.2 Tkinter GUI ... 137
8.5 Reading a PIR Sensor ... 137
8.5.1 Recipe to Read PIR Sensor ... 137
8.5.2 Tkinter GUI ... 140
8.6 Reading an Analog Sensor ... 140
8.6.1 Recipe ... 141
8.6.2 Tkinter GUI ... 143

9. Data Acquisition with Python and Tkinter ... 145
9.1 Basics ... 145
9.2 CSV File ... 146
9.3 Storing Arduino Data with CSV File ... 147
9.3.1 Recipe ... 148
9.4 Plotting Random Numbers Using Matplotlib ... 149
9.4.1 Recipe ... 150
9.5 Plotting Real-Time from Arduino ... 150
9.5.1 Recipe ... 151
9.6 Integrating the Plots in the Tkinter Window ... 152
9.6.1 Recipe ... 152

Section IV Connecting to the Cloud

10. Smart IoT Systems ... 157
10.1 DHT11 Data Logger with ThingSpeak Server ... 157
10.1.1 Installation of DHT11 Library ... 157
10.1.2 Steps to Create a Channel in ThingSpeak ... 157
10.1.3 Recipe ... 160
10.2 Ultrasonic Sensor Data Logger with ThingSpeak Server ... 160
10.2.1 Recipe ... 161
10.3 Air Quality Monitoring System and Data Logger with ThingSpeak Server ... 162
10.3.1 Recipe ... 166
10.4 Landslide Detection and Disaster Management System ... 169
10.4.1 Recipe ... 171
10.5 Smart Motion Detector and Upload Image to gmail.com ... 173
10.5.1 Configuring Raspberry Pi with Camera and Gmail ... 174
10.5.2 Recipe ... 175

11. Blynk Application with Raspberry Pi 177
11.1 Introduction to Blynk 177
11.1.1 Installing Blynk on Raspberry Pi 177
11.2 Creating New Project with Blynk 178
11.3 Home Appliance Control with Blynk App 180

12. Cayenne Application with Raspberry Pi 183
12.1 Introduction to Cayenne 183
12.1.1 Getting Started with Cayenne 183
12.2 LED Blynk with the Cayenne App 183

Bibliography 187

Index 189

Preface

The Internet of Things provides direct integration of the physical world with computer-based systems, which results in economic benefits, efficiency improvements, and reduced human exertions.

The primary objective of writing this book is to provide a platform to the readers to get started with the Internet of Things with Raspberry Pi and Arduino along with the basic knowledge of programming and interfacing of the input/output devices.

The book provides a gradual pace of basic to advanced interfacing and programming of Arduino and Raspberry Pi. It is comprised of 4 sections and 12 chapters with a few case studies to make the concept clear and test its feasibility in solutions to real-time problems.

This book is entirely based on the practical experience of the authors while undergoing projects with students and industries.

Acknowledgments

We acknowledge the support from nuttyengineer.com, who provided its products in order to demonstrate and explain how the systems worked. We would like to thank the publisher for encouraging our ideas in this book and their support in helping us to manage this project efficiently.

We are grateful to the honorable chancellor (Lovely Professional University) Ashok Mittal, Mrs. Rashmi Mittal (pro chancellor, LPU), and Dr. Ramesh Kanwar (vice chancellor, LPU), for their support and constant encouragement. In addition, we are thankful to our family, friends, relatives, colleagues, and students for their moral support and blessings.

Dr. Rajesh Singh

Dr. Anita Gehlot

Dr. Lovi Raj Gupta

Mr. Bhupendra Singh

Mr. Mahendra Swain

Authors

Dr. Rajesh Singh is currently associated with Lovely Professional University as a professor with more than 15 years of experience in academics. He has received a gold medal in MTech and honors in his BE degree. His areas of expertise include embedded systems, robotics, wireless sensor networks, and Internet of Things. He has organized and conducted a number of workshops, summer internships, and expert lectures for students as well as faculty. He has 23 **patents** in his account. He has published approximately 100 **research papers** in referred journals/conferences. He has been invited as the session chair and keynote speaker to international/national conferences and faculty development programs.

Under his mentorship, students have participated in national/international competitions including a Texas competition in Delhi and a Laureate Award of Excellence in robotics engineering in Spain. Twice in the last four years he has been awarded the **Certificate of Appreciation** and the **Best Researcher Award 2017** from the University of Petroleum and Energy Studies for exemplary work. He received the **Certificate of Appreciation** for mentoring the projects submitted to the Texas Instruments Innovation Challenge India Design Contest from Texas Instruments in 2015. He has been honored with the **Certificate of Appreciation** from Rashtryapati Bhawan for a mentoring project through the Gandhian Young Technological Innovation Award 2018. He has been awarded the Young Investigator Award at the International Conference on Science and Information in 2012. He has published 15 books in the areas of embedded systems and Internet of Things with reputed publishers like CRC/Taylor & Francis, Bentham Science, Narosa, GBS, IRP, NIPA, River Publisher, and RI Publications. He is editor to a special issue published by the AISC book series at Springer with title *Intelligent Communication, Control and Devices*, 2017–2018.

Dr. Anita Gehlot is currently associated with Lovely Professional University as an associate professor with more than 10 years of experience in academics. She has **20 patents** in her account. She has published more than **50 research papers** in referred journals and conferences. She has organized a number of workshops, summer internships, and expert lectures for students. She has been invited as session chair and keynote speaker to international/national conferences and faculty development programs.

She has been awarded with the **Certificate of Appreciation** from the University of Petroleum and Energy Studies for exemplary work. She has been honored with the **Certificate of Appreciation** from Rashtryapati Bhawan for a mentoring project through the Gandhian Young Technological Innovation Award 2018. She has published 15 books in the areas of embedded systems and Internet of Things with reputed publishers like CRC/Taylor & Francis Group, Bentham Science, Narosa, GBS, IRP, NIPA, River Publisher, and RI Publications. She is an editor to a special issue published by the AISC book series at Springer with the title *Intelligent Communication, Control and Devices*, 2018.

Dr. Lovi Raj Gupta is the executive dean of the faculty of Technology & Sciences at Lovely Professional University. He is a leading light in the field of technical and higher education in the country. His research-focused approach and insightful innovative intervention of technology in education has won him much accolades and laurels.

He holds a PhD in bioinformatics. He received his M.Tech. in computer-aided design and interactive graphics from IIT, Kanpur, and received his BE (Hons.) from MITS, Gwalior. Having a flair for endless learning, he has received more than 20 certifications and specializations online on Internet of Things (IoT), augmented reality, and gamification from the University of California at Irvine, Yonsei University, South Korea, Wharton School, University of Pennsylvania, and Google Machine Learning Group. His research interests are in the areas of robotics, mechatronics, bioinformatics, IoT, AI, and ML using tensor flow (CMLE) and gamification.

In 2001, he was appointed assistant controller (technology), ministry of IT, by the government of India by the honorable president of India in the Office of the Controller of Certifying Authorities (CCA). In 2013, he was awarded a role in the National Advisory Board for the What Can I Give Mission—Kalam Foundation of Dr. APJ Abdul Kalam. In 2011, he received the MIT Technology Review Grand Challenge Award followed by the coveted Infosys InfyMakers Award in 2016. He has **10 patents** to his account.

Bhupendra Singh is the managing director of Schematics Microelectronics and provides product design and R&D support to industries and universities. He has completed his BCA, PGDCA, M.Sc. (CS), and M.Tech and has more than 11 years of experience in the fields of computer networking and embedded systems. He has published 12 books in the areas of embedded systems and Internet of Things with reputed publishers like CRC/Taylor & Francis, Bentham Science, Narosa, GBS, IRP, NIPA, River Publisher, and RI Publications.

Mahendra Swain is presently associated with MIET, Jammu, as assistant professor. He has completed his BTech in ECE from Centurion University of Technology and Management, Bhubaneswar, and M.Tech from Lovely Professional University. He has published various research articles and attended national and international conferences in the fields of embedded systems and Internet of Things. He has chosen electrical and communication engineering with an intense urge to delve into challenging fields, such as embedded systems, wireless communication, electronic product development, and IoT.

Section I

Introduction

1

Introduction to Internet of Things

1.1 Characteristics of IoT

Internet of Things (IoT) can be used to design products for businesses. It facilities to add the valuable feature to the businesses, where the IoT framework is designed to connect the information from devices which are interconnected. The process has been classified into five phases. The first phase is the "create phase," where sensors collect the data from the environment. This data can generate the information for the business. Second is "communicate phase," where data generated in the first phase are communicated to the required destination. Third is the "aggregate phase," where collected data over the network are aggregated by the device itself. Fourth is "analyse phase," where aggregated data are used to generate patterns and use it to control and optimize the process. Fifth is the "act phase," where necessary actions are taken on the basis of information extracted from the aggregated data.

The characteristics of IoT may vary from one domain to other. A few characteristics are listed as follows:

1. **Intelligence:** IoT is treated as smart due to the integration of hardware, software, computational capability, and algorithms. The intelligence feature in an IoT system establishes a great capability of responding in an intelligent way to situations in order to carry out specific tasks. IoT provides a standard input method in the form of a graphical user interface, which also makes it user friendly.
2. **Connectivity:** Connectivity brings objects together over IoT. It is important, as connectivity contributes to the collective intelligence of the system. It enables network accessibility and compatibility of the objects. New opportunities could be generated in the recent market by connecting smart devices by the networking.
3. **Dynamic nature:** IoT is dynamic in nature, as it is capable of collecting data from devices, which may change dynamically, for example, a change in temperature or speed.

4. **Enormous scale:** The number of connected devices over IoT is very large. The data management for such large number of devices is even more critical. But the complexity does not affect the number of objects connected to IoT day by day.
5. **Sensing:** The sensor is the most important part of an IoT network. It detects or measures the environmental changes to generate the data. The sensing technologies provide true information regarding physical quantities in the environment.
6. **Heterogeneity:** IoT devices designed using various hardware frameworks and networks can communicate over different networks. Features like modularity, scalability, extensibility, and interoperability play key design roles in IoT.
7. **Security:** IoT devices are susceptible to cyberattack. There is a high level of transparency and privacy issues with IoT. It is important to secure the end objects, the networks, and the data that are transferred over the network.

There is a wide range of technologies incorporated with IoT to support its successful functioning. IoT creates values and supports human beings to make their lives better.

1.2 Design Principles of IoT

In the near future, everyday lives will be filled with more intelligent devices. Designing IoT devices and networks has challenges to be addressed, which includes connecting different types of physical devices, collecting data, extracting meaningful information, and fulfilling different needs at the level of industries and home.

A few of the design principles for IoT devices and networks are as follows:

1. **Focus on value:** To start with IoT design, it is important to understand the types of features that need to be included. The challenges and barriers need to be understood before adopting new technologies. The designer needs to dig out the user's need and acceptability of the product. The order of features also needs attention in any design process.
2. **Holistic view:** IoT product consists of multiple devices with different capabilities. The solution may also have a cooperation with multiple service providers. It is not enough to design only one end device; the designer needs to take a holistic view across the complete system.

3. **Safety first:** The consequences of avoiding safety in IoT products can be very serious due to direct connection of devices with the real world. Also, building trust must be one of the main drives among designers. Because IoT is a combination of hardware, software, and network, any situation of occurring error needs to addressed. In the event an error situation can't be avoided, the feature of communicating the error to the user may build a trust. It is important to make users' data secure and safe, to build a trust towards IoT.
4. **Consider the context:** IoT solutions directly deal with the real world, where many unexpected things happen at a time when the user should feel safe. IoT solutions should be capable of handling changing environmental situations, such as a change in temperature. Also, an IoT device can have multiple users, unlike a smartphone, so this context needs to be addressed.
5. **Strong brand:** To handle adverse conditions such as device failure, building a strong brand is very important among users. When users feel connected with a brand, they are more forgiving and are more likely to keep the products.
6. **Prototyping:** An IoT solution is a combination of both hardware and software, and both have different life spans. But in IoT, the solution needs to be aligned. IoT hardware and software are hard to upgrade once they are placed at a location. So, the prototyping and its iteration are the solution before actual finalizing the product to launch.
7. **Use data responsibly:** An IoT solution generates tons of data during its life span. However, the idea is not to hold all the data but instead to identify the data points that are required to make the solution functional and useful. So, the possibility of data sciences comes in here. Data science provides a solution to reduce user friction. It can be used to interpret meaningful signals and automate repeated context-dependent decisions.

1.3 IoT Architecture and Protocols

1.3.1 IoT Architecture

IoT architecture is comprised of the following components:

1. **Thing:** IoT is interconnected with various sensors to collect the data and actuators to perform actions corresponding to the commands received from the cloud.
2. **Gateway:** It is used for data filtering, preprocessing, and communicating it to the cloud and vice versa (receiving the commands from the cloud).

3. **Cloud gateway:** It is used to transmit data between the gateways and IoT central servers.
4. **Streaming data processor:** It distributes the data coming from sensors to the relevant devices connected in network.
5. **Data lake:** It is used to store all defined and nondefined data.
6. **Big data warehouse:** It is used for collecting valuable data.
7. **Control application:** It is used to send commands to the actuators.
8. **Machine learning:** It is used to generate models by applying algorithms on data, which can be used to control applications.
9. **User application:** It enables the users to monitor the data and make decisions on controlling connected devices.
10. **Data analytics:** It is used for manual data processing.

1.3.2 IoT Protocols

1.3.2.1 OSI Model

The OSI (Open Systems Interconnection) model for IoT protocols, as shown in Figure 1.1, includes five layers: physical layer, link layer, internet layer, transport layer, and application layer.

The physical layer is comprised of devices, objects, and things. The link layer operates on protocols like IEEE 802.15.4, IEEE 802.11, IS/IEC 18092:2004, Bluetooth, ANT, NB-IoT, EC-GSM-IoT, ISA100.11a, EnOcean, and LTE-MTC. The internet layer protocols are 6LoWPAN, IPv6, uIP, and NanoIP. The transport layer protocols are CoAP, TCP, UDP, MQTT, XMPP, AMQP, LLAP, DDS, SOAP, and DTLS. The application protocols are JSON-IPSO, REST API objects, and binary objects.

Layer	Protocols
Application layer	RES TAPI, JSON-IPSO objects, Binary objects
Transport layer	CoAP, MQTT, XMPP, AMQP, LLAP, DDS, SOAP, UDP, TCP, DTLS
Internet layer	6LoWPAN, IPv6, uIP, NanoIP
Link layer	IEEE802.15.4, IEEE802.11, ISO/IEC 8092:2004, NB-IoT, EC-GSM-IoT, Bluetooth, ANT, ISA100.11a, EnOcean, LTE-MTC
Physical layer	Devices, objects, things

FIGURE 1.1
OSI model for IoT protocols.

1.3.2.2 Organizational Levels

IoT protocols can also be categorized on the basis of the organization levels, as follows:

1. Infrastructure (IPv4/IPv6, 6LowPAN, RPL)
2. Identification (EPC, IPv6, uCode, URIs)
3. Communication (Bluetooth, Wi-Fi, LPWAN)
4. Discovery (DNS-SD, mDNS, Physical Web)
5. Data Protocols (AMQP, MQTT, Websocket, CoAP, Node)
6. Device Management (TR-069, OMA-DM)
7. Semantic (Web Thing Model, JSON-LD)
8. Multi-layer Frameworks (Weave, IoTivity, Alljoyn, Homekit)

IPv6: IPv6 is popular as an internet layer protocol to transmit packets of information in an end-to-end transmission over multiple Internet Protocol (IP) networks.

6LoWPAN: 6LoWPAN stands for IPv6 over Low-power Wireless Personal Area Networks. It is an extended layer for IPv6 through IEEE802.15.4 links. It operates on 2.4 GHz of frequency with a data transmission rate of 250 kbps.

RPL: It is an IPv6-based routing protocol used in less power and lossy network.

UDP (User Datagram Protocol): This protocol is meant for IP-based protocol networked between client/server. UDP is used in applications for real-time performance.

QUIC: It stands for Quick UDP Internet Connections. It supports the multiplexed connections between two endpoints over (UDP). It was designed for security protection to reduce latency in connection and transportation of information over a network.

μIP: The acronym is micro Internet Protocol. It is widely used due to open source TCP/IP stack, which can be used for tiny 8- and 16-bit microcontrollers.

DTLS (Datagram Transport Layer): The DTLS protocol provides the communication privacy for datagram protocols. It is used for prevention of tampering, message forgery, or eavesdropping in a network.

NanoIP: Nano Internet Protocol is meant to establish communication among embedded sensors and devices without the overhead of TCP/IP.

Time-Synchronized Mesh Protocol (TSMP): It is a communication protocol to establish communication among self-customized wireless sensor nodes called motes.

Physical Web: The Physical Web is an approach to interconnect devices and access them seamlessly.

HyperCat: It is an open source JSON-based lightweight hypermedia catalogue format for exposing collections of URIs.

MQTT (Message Queuing Telemetry Transport)**:** The MQTT is a lightweight protocol that enables a publish/subscribe messaging model. Used for remote connection in a network.

CoAP (Constrained Application Protocol): CoAP is an application layer protocol. It is designed to translate to HTTP for simplified integration with the web.

SMCP: It is a C-based CoAP stack, which can be used for embedded environments. It has fully asynchronous I/O and supports both UIP and BSD sockets.

STOMP: It stands for Simple Text Oriented Messaging Protocol used in communication networks.

XMPP: It stands for Extensible Messaging and Presence Protocol.

XMPP-IoT: It is the same as XMPP with an additional feature to establish a communication link between machine to people and machine to machine.

Mihini/M3DA: It acts as intermediate agent between the M2M server and embedded gateway. M3DA is an extended version to transport M2M binary data.

AMQP: The acronym is Advanced Message Queuing Protocol and is an open-source application layer used as middleware in a messaging application. It is reliable and more secure to use in routing and queuing.

DDS: It stands for Data Distribution Service for real-time systems. It is an open source and international standard to address communication among real time and embedded systems.

LLAP: It is elaborated as a lightweight local automation protocol. LLAP facilitates sending short and simple messages between intelligent objects.

REST: It stands for Representational State Transfer.

SOAP: It stands for Simple Object Access Protocol.

Websocket: It is a full-duplex socket used to communicate between server and client.

SensorML: It describes sensors and the measurement process by providing the standard models and an XML encoding.

RAML: The acronym is RESTful API Modelling Language. It is used to design and share the API.

IoTivity: It is founded by Linux Foundation to facilitate open-source projects and sponsored by OIC.

IEEE P2413: It is a standard for an architectural framework for the IoT.

OTrP (Open Trust Protocol): This protocol is used to install, update, and delete applications. It manages the security configuration in a Trusted Execution Environment (TEE).

1.4 Enabling Technologies for IoT

In the present world, many wired and wireless technologies contribute to automation. IoT is the latest trend in technology. The networking part in IoT may involve more than one type of communication media or device.

1. **Short-Range Wireless Technology**

   Bluetooth networking in mesh: It is a Bluetooth low-energy (BLE) compatible mesh network with an increased number of nodes.

   Light-Fidelity (Li-Fi): This technology is almost similar to the Wi-Fi standard, but it uses the visible light spectrum.

   Near-field communication (NFC): It is a communication protocol that enables communication between two devices within a range of 4 cm.

   QR codes and barcodes: It is optical tag that can be read by machine; it stores the information for the item to which it is stacked to.

   Radio-frequency identification (RFID): It uses electromagnetic fields to read the information stored in tags on the other items.

   Thread: This network protocol is based on the IEEE 802.15.4 standard.

   Wi-Fi: It is for local area networking, which is based on the IEEE 802.11 standard.

   Z-Wave: It is a low-powered, low-latency, near-range communication protocol having better reliability than Wi-Fi.

   ZigBee: This protocol can be used for a personal area network; it is based on the IEEE 802.15.4 standard.

2. **Medium-Range Wireless Technology**

   HaLow: It is the variant of the Wi-Fi standard. It provides low data rate transmission over a wide range.

   LTE-Advanced: It is Long-Term Evolution technology meant to provide flawless communication with a high data rate.

3. **Long-Range Wireless Technology**

   Low-power wide-area networking (LPWAN): This wireless network facilitates a wide range of communication along with low bit rate and less power.

   Very small aperture terminal (VSAT): This communication is used in satellites using dish antenna for narrow-banded data.

4. **Wired Technology**

   Ethernet: It is a wired communication technique using a twisted pair and optical fiber with hubs or switches.

   Multimedia over Coax Alliance (MoCA): This technology enhances video quality over existing cable.

   Power-line communication (PLC): This communication technology uses the transmit of electrical power and data.

1.5 IoT Levels

Level 1 IoT: A level 1 IoT system performs sensing, actuation, storing, and analysis operations and is comprised of a single node/device. An example is a home automation system where a single node is designed to control the lights and appliances remotely.

Level 2 IoT: A level 2 IoT system performs sensing, actuation, and analysis and has a single node/device. This is suitable for big data analysis. The data is stored on the cloud. It is popular for cloud-enabled applications like smart farming.

Level 3 IoT: A level 3 IoT system is a single-node-based cloud platform. This type of system is suitable for big data needs that are computationally intensive. An example is the package tracking system. The system comprises of a single node (for a package), which monitors the vibration level of a package being shipped.

Level 4 IoT: A level 4 IoT system has multiple nodes that perform the analysis and data stored on the cloud. The system may have local and cloud-based server nodes that receive the information and upload on the cloud. Server nodes only process the information and perform no control action. It is suitable where multiple nodes are required and involve big data that is computationally intensive. An example is noise monitoring.

Level 5 IoT: A level 5 IoT system has multiple end nodes and a single coordinator node. The end node performs the sensing and/or actuation actions. Collections of data done by the coordinator node form the sensor nodes and communicates it to the cloud and is analyzed on the cloud. The system is suitable for a WSN-based solution with big data and computationally intensive requirement. An example is forest fire detection. The system is comprised of multiple nodes placed at different locations for monitoring temperature, humidity, and CO_2 levels in the forest.

Level 6 IoT: A level 6 IoT is comprised of sensor nodes and an actuator to perform sensing and controlling. It is a suitable cloud-based database designed for data analysis. The central controller knows the status of all end nodes and sends the control commands to the nodes. An example is a weather monitoring system. The system is comprised of multiple nodes that are placed at the different locations for monitoring temperature, humidity pressure, radiation, and wind speed. The sensor nodes are responsible for transmission of the data from end nodes to the destination via a websocket. The data is stored on the cloud-based server. The data analysis is done on the cloud to make the prediction by aggregating the data.

1.6 IoT vs M2M

IoT can be defined as a system where multiple objects communicate with each other and share data through sensors and digital connectivity. Machine-to-machine (M2M) solutions are comprised of linear communication channels between the machines to make them work in a cycle. Here, the action of one machine triggers the activity of other.

Differences between IoT and M2M

- A few experts define M2M as a subset of IoT, while others call the Internet of Things an evolved version of machine to machine. Either way, the conclusion is IoT is a broader area than M2M.
- Both the technologies work on the principle of connecting devices and make them to work together. While M2M relies on conventional connection tools like Wi-Fi, IoT has much flexibility and varied connectivity options.
- M2M solution has very limited scope and is confined to create a network of machines that work in synchronization. IoT creates 360° solutions for flexible responses and multi-level communication.
- The advantage of IoT over M2M is its ability to add interactivity amongst devices. Machine to machine operates by triggering responses based on an action. It is a one-way communication. In IoT-based systems, communication flows to and fro freely.

2

Sensors

2.1 Sensors Classification

Sensors are classified as follows:

1. Primary input quantity
2. Transduction principles
3. Technology and material
4. Property
5. Application

Classification of sensors can also be done on the basis of different areas:

1. **Classification based on application:** Sensors are chosen on the basis of application where they need to be implemented, such as industrial process control, measurement and automation, automobiles, consumer electronics, aircraft, and medical products. As with the change in application, the selection criteria changes, so the application needs to be considered.
2. **Classification based on power or energy supply requirement**

   Active sensor: Active sensors are those where a power supply is required to measure the physical quantity, e.g., temperature sensor, ultrasonic sensor, and light-dependent resistor (LDR).

   Passive sensor: Sensors that do not need a power supply are called passive sensors, and they measure the parameters, e.g., radiometers film photography.
3. **Classification based on output of sensor**

   Digital sensor: The output of sensor is in binary or digital form, which can be directly processed through a controller or processor.

   Analog sensors: The output of a sensor is in the form of a continuous signal. An analog-to-digital converter is required to read the sensor by microcontroller or processor.

4. **Classification based on the type of sensor:** There are several sensors available with different applications. Sensors can be categorized on the basis of types of sensors. A few types of sensors are discussed as follows:

   Accelerometers: Accelerometers are based on the technology named "microelectromechanical sensor." They can be used in dynamic systems.

   Biosensors: Biosensors are based on the electrochemical technology. They can be used for medical care devices, water testing, food testing, etc.

   Image sensors: These are developed on the basics of the complementary metal oxide semiconductor (CMOS) technique. These are widely used in to video surveillance, biometrics, and traffic management.

   Motion detectors: Motion detectors are based on the infrared, ultrasonic, and microwave/radar technology. These are used in security purposes.

5. **Classification based on property:** The sensors are also classified on the basis of the property of the physical parameter. A few examples are as follows:

   Temperature: Thermocouples, thermistors, resistance temperature detectors (RTDs)

   Flow: Thermal mass, differential pressure, electromagnetic, positional displacement, etc.

   Pressure: Fiber optic, linear variable differential transformer (LVDT), elastic liquid-based manometers, vacuum, electronic

   Level sensors: Ultrasonic radio frequency, radar, thermal displacement, etc.

   Proximity and displacement: Capacitive, LVDT, magnetic, photoelectric, ultrasonic

   Biosensors: Electrochemical, resonant mirror, surface plasmon resonance

   Image: Charge-coupled devices, CMOS

   Gas and chemical: Semiconductor, conductance, infrared, electrochemical

   Acceleration: Accelerometers, gyroscopes

2.2 Working Principle of Sensors

The working principle of each sensor is different, as it is designed to measure a specific quantity. The principle of few basic sensors is as follows:

1. **Temperature sensor:** The temperature sensor measures the environmental temperature and converts it to an electrical signal. The principle of the thermometer is expansion and contraction of mercury in glass. With an alteration in temperature, mercury expands and contracts proportionally.

   Two types of temperature sensors are available:

   Contact sensor: The sensor that needs to be in physical contact with the object, temperature of which is to be sensed, is known as a contact sensor.

   Noncontact sensor: The sensor that needn't to be in physical contact with the object, temperature of which is to be sensed, is known as a noncontact sensor. This type of sensor uses Plank's Law to measure temperature, which senses the heat radiated from the source to measure the temperature.

   Examples of temperature sensors:

   Thermocouple: Thermocouple is made of two wires, each with different metals. A junction is formed by joining the ends. This junction is open to the object for which temperature needs to be measured; the other end is connected to a measuring device. The current will flow through the metal, due to a difference in temperature of two junctions.

   Resistance temperature detectors (RTDs): An RTD is type of thermal resistor that is designed to alter the electrical resistance with a change in temperature.

   Thermistors: It is type of thermal resistor that changes the resistance in proportion with small changes in temperature.

2. **IR sensor:** An IR sensor emits and detects the infrared rays to sense a specific environment. It is easily available in the market, but it is sensitive toward noise and light.

   The application of an IR sensor includes thermography, heating, meteorology, climatology, spectroscopy, and communications.

3. **UV sensor:** A UV sensor measures the intensity or the power of an incident ultraviolet radiation. This electromagnetic radiation has longer a wavelength than x-rays but smaller than visible radiation. A polycrystalline diamond material is used for ultraviolet sensing. It can transmit different types of energy signals but can accept only one type of signal. The electrical meter is used to read the output signals and processed to the computer through analog-to-digital converters. The UV sensor is used in UV water treatment, light sensors, UV spectrum detectors, etc.
4. **Touch sensor:** A touch sensor is a variable resistor that changes its resistance as per the location where it gets touched. It is made of a conductive and a partially conductive substance and insulated in a plastic cover. The flow of current is due to a conductive material that allows current partially. The touch sensor is a cost-effective solution for many applications, such as washing machines, fluid-level sensors, and dishwashers.
5. **Proximity sensor:** A proximity sensor can detect the presence of an object without any contact point. The working principle is electromagnetic waves that are emitted by the sensor and return when the object is in range of the waves. The presence of the object is detected with the change in filed radiation. The proximity sensors working are of different types, like inductive, capacitive, photoelectric sensor, ultrasonic, and Hall-effect.

   Inductive proximity sensor: This type of sensor has an oscillator as an input, which changes the loss resistance by the proximity of an electrically conductive medium. For metal detection, these types of sensors are used.

   Capacitive proximity sensor: This type of sensor converts capacitance by changing electrode displacement. It can be done by bringing the object within the variable frequencies. The object is detected with the help of the oscillated frequency, which is converted into a DC voltage. This current is compared with a fixed value to detect the object. For plastic targets, these types of sensors are used.
6. **Ultrasonic sensor:** An ultrasonic sensor is used to detect the distance of an object. The working principle is the time duration between the emission and receiving of the waves after reflecting from the object. Ultrasonic sensors use sound waves to measure the distance of an object.

2.3 Criteria to Choose a Sensor

There are a few features that need to be addressed, along with the sensor to be selected. The features are as follows:

1. Accuracy
2. Cost
3. Range of communication
4. Repeatability
5. Resolution
6. Environmental constraints
7. Data calibration

2.4 Generation of Sensors

First generation: The first-generation sensors were associated with electronics. Most of the structures were based on silicon structure. Few sensors had the facility of analog amplification on a microchip.

Second generation: This generation of sensors was analog in nature with MEMS element combined with analog amplification. These had the facility of an analog-to-digital converter on one microchip.

Third generation: This generation of sensors had a combination of sensor element, analog amplification, and analog-to-digital converter with the on-chip digital intelligence and temperature compensation.

Fourth generation: This generation of sensors had an additional feature of memory cell for calibration and temperature compensation, along with the features of the third generation.

Fifth generation: This is generation of intelligent sensors with the capability of communication.

3

IoT Design Methodology

3.1 Design Methodology

The process of developing a product needs a proper execution of predefined design methodology. Generally, a product development process includes the following steps:

1. Describe the objective
2. Requirements to achieve the objective
3. Design the system architecture
4. Identify the stages of development
5. Assembling and coding each stage
6. Integrate all stages
7. Testing and troubleshooting
8. Debugging
9. Launch of product

Internet of Things (IoT) system design is the complete design of devices that are interconnected with each other. It combines physical and digital, both components to collect data from remote devices and deliver actionable signals.

It consists of various sensors, secured communication, gateway, cloud-enabled servers, and dashboards. All these components need to be designed with consideration of their interdependency.

The basic design principles are as follows:

Interoperability: It is the ability of a system to exchange the information and make use of it. The basic requirement components are sensor, machine, and device to communicate.

Information transparency: The connected devices over a network share the information. Information transparency is recording of the physical process and storing it virtually.

Technical assistance: Technical assistance is the ability of providing and displaying data of a connected system. It is to solve the issue and make the operational decisions easier, to improve the productability.

Decentralized decision: The decision-making is the principle for the connected system; it is required to execute the process with defined logic.

3.2 Challenges in IoT Design

IoT is a combination of many domains at a single platform, so the designing of an IoT system is quite challenging. The data security of customers is a big challenge for the service providers. If an IoT device suffers with connectivity problems due to a poor network, the purpose of deploying IoT is futile. It will be an even bigger issue when a large number of devices are connected. Heterogeneity of the network has different challenges with respect to security, privacy, and functionality.

A few of the challenges of IoT design are as follows:

1. **Availability:** Availability is the consistency of the network, even in case of attacks. As IoT services need to be real time, so security with availability is the prime concern.
2. **Authenticity:** It is the process where users need to prove the identity to access the services. It is required for the protection of the system. It is required to avoid illegal access of services.
3. **Confidentiality:** For data confidentiality, only an authorized person can access or modify the data.
4. **Integrity:** The data received by the user is undamaged, unmodified, and original as sent by the sender; this assurance is provided by integrity.
5. **Nonrepudiation:** It is the assurance of sending the correct information by end node without denying the sharing of data at any time and the acknowledgment from the receiver for the same.

3.3 IoT System Management

According to a study by International data corporation (IDC), the number of devices connected to the internet is expected to reach 30 billion by 2020. Management is the most important part of any system. The IoT system management includes the device deployment, provision of the device and authentication, configuration and control, monitoring and diagnostic, and then software updates and maintenance. IoT is not only about deploying the sensors and capturing the data to communicate to the server, but once the system is established, there may be a requirement of software updates, and repairing and replacing the faulty devices along with the security of data.

Provisioning and authentication: Authentication is a process of establishing the identity to ensure the security and trust. A cloud-hosted service is required to be implemented to check the authenticity of the software and hardware connected to the network.

Provisioning is a method to provide access of a device to a system with suitable authentication.

Configuration and control: Configuring a system means an arrangement of parts in a specific form, figure, or combination of elements. Configuring an IoT device includes attributes, such as name, location, and specific settings for application.

IoT devices need to be configured and authenticated from user attributes to make it reliable. The control is capability of handling the device and help for configuration changes.

Monitoring and diagnostics: Monitoring is the process of observing the progress of a system over a time period. The IoT system is connected with thousands of remote devices over the internet, and a small mistake in data monitoring may cause loss of trust of the customer. Even small issues need to be addressed with appropriate diagnosis of the problem. For troubleshooting, the developers need to implement good a program and must be capable of updating through cloud analytics.

Software maintenance and update: Software maintenance is another task in IoT that needs to support firmware, which should be free from any bugs. However, updating firmware is another important concern. The developer must have secure updated software, including the boot loaders.

Software maintenance on remote devices is a long-term process. A persistent and reliable connection is required with remote devices for maintenance and updating. This is a complicated process and needs to be performed when there is minimum impact on the business.

3.4 IoT Servers

Many cloud providers are in the market, which provides suitable IoT-based services for specific applications.

3.4.1 KAA

KAA is an IoT middleware platform and open-source framework for building smart connections for end-to-end IoT solutions with Apache License 2.0. It provides services for data analysis, visualization, and cloud service for IoT systems (http://www.kaaproject.org/).

3.4.2 SeeControl IoT

SeeControl is an IoT-enabled cloud platform that is efficient in data analytics and data visualization to maintain proper work flow in monitoring and controlling (http://www.seecontrol.com).

3.4.3 Temboo

Temboo is a cloud-based platform for application code generation. It involves less wiring and coding of hardware and software. It has more than 90 inbuilt libraries named "Choreos" for third-party services, including Yahoo weather, Twilio telephony, eBay product shopping, Flickr photo management, Amazon cloud, Twitter microblogging, Facebook Graph API, Google analytics, PayPal payment, Uber vehicle confirmation, YouTube video streaming, and many more (https://temboo.com).

3.4.4 SensorCloud

SensorCloud is an IoT cloud that provides the Platform as a Service (PasS) to gather, visualize, monitor, and analyze the information coming into sensors connected by wire or wirelessly. It allows the data to be analyzed with complex mathematical algorithms (http://www.sensorcloud.com).

3.4.5 Carriots

Carriots is platform that helps anyone to build quick IoT applications. It saves time, cost, and troubles. The PasS is designed to add features like remote device management and control, rule-based listeners' activity logging, triggering custom alarms, and data export (https://carriots.com).

3.4.6 Xively

Xively is a gravity cloud technology-based IoT cloud service. It helps companies manage their products by addressing various features like scalability, reliability, and secure. It is easy to integrate with devices but has minimum notification services (https://xively.com).

3.4.7 Etherios

Etherios supports comprehensive products and services for connected enterprises. Its cloud is designed on the PaaS model to enable users for connecting product and gain real-time visibility into their assets. It is a specialized cloud, but developers are restricted with limited devices (http://www.etherios.com).

3.4.8 thethings.io

thethings.io is a platform that gives a complete solution for the back-end developer with easy and flexible application program interfaces (APIs). thethings.io is a hardware agnostic that allows the connection of any device that is capable of using MQTT, CoAP protocols, HTTP, or WebSockets (https://thethings.io).

3.4.9 Ayla's IoT Cloud Fabric

Ayla IoT fabric is a PaaS-modeled enterprise class. Ayla Networks provide a firmware intermediator embedded in both devices and mobile device applications for end-to-end support. It provides easy mobile application development platform but is not suitable for small-scale developers (https://www.aylanetworks.com).

3.4.10 Exosite

Exosite is modular, enterprise-grade IoT software platform that helps to bring connected products to the market. It has a cloud platform based on IoT Software as a Service (SaaS), which provides real-time data visualization and analytics support to the users. The system development is easy with it, but it lacks in big data provisions (https://exosite.com).

3.4.11 OpenRemote

OpenRemote is an open source the IoT middleware solution, which allows users to integrate any device—protocol—design using available resources like iOS, Android, or web browsers. It supports open cloud services but has a high cost (http://www.openremote.com).

3.4.12 Arrayent Connect TM

Arrayent is an IoT platform that enables heterogeneous brands like Whirlpool, Maytag, and First Alert to connect users' products to smart handheld devices and web applications. Arrayent Connect Cloud is an IoT operating system that is based on the SaaS model (http://www.arrayent.com).

3.4.13 Arkessa

Arkessa provides services to companies to empower them with maximum revenue and to enhance customer satisfaction. It helps companies to develop IoT devices to enhance connectivity, monitoring, and controlling with enterprise. It has enterprise-enabled design facets, but its visualization apps are not proper (http://www.arkessa.com).

3.4.14 Oracle IoT Cloud

It is comprised four crucial parameters. It performs operations on received data including analysis, acquisition, and integration. It supports the database but lacks in connectivity of open-source devices (https://cloud.oracle.com/iot).

3.4.15 ThingWorx

ThingWorx is a data-driven decision-making cloud. It provides M2M and IoT services based on SQUEAL. Zero coding facility is available (https://thingworx.com).

3.4.16 Nimbits

Nimbits is a cloud server that provides solutions to edge-computing IoT-related services. It performs operations such as noise filtering and sends data on the cloud. It is easy to adopt but lacking in the real-time processing of query (http://www.nimbits.com).

3.4.17 InfoBright

InfoBright is an IoT-based analytical database platform that connects business to store and acts on machine-generated data for a complete ecosystem (https://www.infobright.com/index.php/internet-of-things).

3.4.18 Jasper Control Center

Jasper Control Center is a platform based on Jasper control. Control center is designed to automate the connected devices and help to analyze real-time behavior patterns. The main advantage is its rule-based behavior pattern (https://www.jasper.com).

3.4.19 AerCloud

AerCloud platform collects, manages, and analyzes sensory data for IoT and M2M applications. It is scalable to M2M services but not suitable for developers (http://www.aeris.com).

3.4.20 Echelon

Echelon is an IoT-based platform for the cloud with resources like microphones, hardware devices, and other applications. It is good for the industrial prospective but lacks in basics for beginners (http://www.iiot.echelon.com).

3.4.21 ThingSpeak

It is an open-source public cloud platform specially developed for IoT-based applications. It has open API that receives real-time data. It has data storage, monitoring, and visualization facilities (https://thingspeak.com).

3.4.22 Plotly

Plotly is a data visualization cloud service provider for the public. It provides data storage, analysis, and visualization services. Python, R, MATLAB, and Julia-based APIs are implemented in Plotly (https://plot.ly).

3.4.23 GroveStreams

GroveStreams is a public cloud for data visualization. It supports various data types. It enables seamless monitoring but lacks in statistical services (https://thingworx.com).

3.4.24 IBM IoT

IBM IoT is an organized architecture cloud platform. It supports complex industry solutions. It can enable a device's identity but the application prototyping is difficult (https://internetofthings.ibmcloud.com).

3.4.25 Microsoft Research Lab of Things

Lab of Things is an IoT platform design developed by Microsoft. It is used to analyze experimental research evidences in academic institutions (http://www.lab-of-things.com).

3.4.26 Blynk

It is an open-source platform with iOS and Android apps, which allows the control of Raspberry Pi and Arduino over the internet. It supports a graphical interface to build projects just by dragging the widgets. It supports many IoT modules.

3.4.27 Cayenne APP

Cayenne is an app for smartphones and computers that controls Raspberry Pi and Arduino through the use of a graphical interface. It has customizable dashboards with drag-and-drop widgets for connection devices. It supports quick and easy setup.

3.4.28 Virtuino APP

Virtuino platform creates amazing virtual screens on smartphones or tablets to control the automation system created with Arduino or similar boards. It supports Arduino and can be connected with the HC-05 Bluetooth, Ethernet Shield, and ESP8266 modules. It supports monitoring sensor values from the IoT server ThingSpeak.

Section II

Basics of Arduino and Raspberry Pi

4

Basics of Arduino

4.1 Introduction to Arduino

Arduino was invented at the Ivrea Interaction Design Institute. It was designed for fast prototyping, targeting the hobbyist without any programming background. Soon the user-friendly platform attracted an audience covering a wider community and started changing to adapt the latest trends in the market, from an 8-bit board to IoT products, wearable devices, and an embedded environment. Arduino boards are completely open source and can be used for application development with particular requirements. The Arduino software is user-friendly and easy to begin with a flexible environment for advanced users. It can be operated on Mac, Linux, and Window platforms. New things can be learned with Arduino.

Advantages of Arduino:

Cost: Arduino boards are less expensive compared to other microcontroller boards.

Platform: The Arduino Software (IDE) is compatible with most of the operating systems like Macintosh OSX, Windows, and Linux.

User friendly: The Arduino Software (IDE) is user-friendly, easy to begin, and has flexibility for the skilled programmers.

Open source: The Arduino is an open-source software that can be programmed with C, C++, or AVR-C languages. So a variety of modules can be designed by users.

4.1.1 Arduino Uno

The Arduino/Genuino Uno has an onboard ATmega328 microcontroller. It has 6 analog input ports (A0–A5) and 14 digital I/O ports, out of which 6 are PWM pins. Each pin can operate on 0–5 V of voltage. It operates at 16 MHz of frequency. Figure 4.1 shows the Arduino Uno board (Table 4.1).

FIGURE 4.1
Arduino Uno board.

TABLE 4.1
Pin Description of Arduino UNO

Pin	Description
Vin	It is the external voltage to the board
3.3 V	3.3 V supply, on board
+5 V	Output voltage +5 V
GND	Ground
IOREF	It is to select the appropriate power source by providing the voltage reference
Serial	It can transmit and receive serial data with 0(Rx) 1(Tx)
External Interrupts	Trigger an interrupt on low value (pins 2 and 3)
PWM	8 bit six PWM (3, 5, 6, 9, 10, 11)
SPI	It supports SPI communication [10 (SS), 11 (MOSI), 12 (MISO) and 13 (SCK)]
LED	Inbuilt LED driven
TWI	TWI communication [A4 (SDA), and A5 (SCL)]
AREF	Reference voltage with the analog inputs
Reset	It is used to reset the onboard microcontroller

4.1.2 Arduino Mega

The Arduino Mega has an onboard ATmega2560 microcontroller. It has 16 analog inputs, 54 digital I/Os, USB connection, 4 UART, a power jack, and a reset button. It operates on 16 MHz frequency. Figure 4.2 shows the Arduino Mega board (Table 4.2).

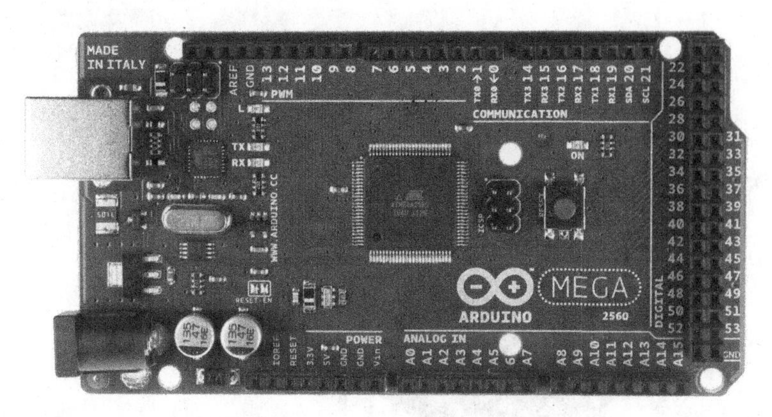

FIGURE 4.2
Arduino Mega board.

TABLE 4.2
Pin Description

Pin	Description
Vin	The external voltage to the Arduino board
+5 V	Output a regulated 5 V
3.3 V	Onboard 3.3 V supply
GND	Ground
IOREF	It is to select the appropriate power source by providing the voltage reference
Serial0	It can transmit and receive serial data with 0(Rx) and 1(Tx)
Serial1	It can transmit and receive serial data with 19(Rx) and 18(Tx)
Serial2	It can transmit and receive serial data with 14(Rx) and 16(Tx)
External Interrupts	It triggers an external interrupt at low value with 2 (interrupt 0), 3 (interrupt 1), 18 (interrupt 5), 19 (interrupt 4), and 20 (interrupt 2)
PWM	8 bit PWM (pins: 2–13 and 44–46)
SPI	It supports SPI communication [10 (SS), 11 (MOSI), 12 (MISO), and 13 (SCK)]
LED	LED driven at pin 13
TWI	Supports TWI communication [pins: 20 (SDA), 21 (SCL)]
AREF	It is a reference voltage for analog inputs
Reset	It is used to reset the microcontroller on board

4.1.3 Arduino Nano

The Arduino/Genuino Nano has an onboard ATmega328 microcontroller. It has 8 analog inputs, 14 digital I/O ports, and 6 PWM. It has onboard 32 KB flash memory, 1 KB EEPROM, 2 KB SRAM, and operates at 16 MHz of frequency. Figure 4.3 shows the Arduino Nano (Tables 4.3 and 4.4).

FIGURE 4.3
Arduino Nano board.

TABLE 4.3
Pin Description of Arduino NANO

PIN	Description
Vin	The external voltage to the board
+5 V	Output as +5 V
3.3 V	3.3 V supply on board
GND	Ground
IOREF	It helps to select the appropriate power source by providing a voltage reference
Serial	It can transmit and receive serial data with 0(Rx) and 1(Tx)
External Interrupts	Trigger an interrupt on low value (pins 2 and 3)
PWM	8 bit PWM (3, 5, 6, 9, 10, 11)
SPI	It supports SPI communication with [10 (SS), 11 (MOSI), 12 (MISO) and 13 (SCK)]
LED	LED driven at pin 13
I2C	Supports two wires interfacing [A4 (SDA) and A5 (SCL)]
AREF	It is a reference voltage for analog inputs
Reset	It is used to reset the microcontroller on board

TABLE 4.4
Comparison Table for a Few Arduino Boards

Name	Processor	CPU Speed	Operating/ Input Voltage	Digital IO/ PWM	Analog In/Out	UART	Flash [kB]
LilyPad	ATmega168V ATmega328P	8 MHz	2.4–5.5 V/ 2.4–5.5 V	14/6	6/0	—	16
Mega 2560	ATmega2560	16 MHz	5 V/4–12 V	54/15	16/0	4	256
Micro	ATmega32U4	16 MHz	5 V/4–12 V	20/4	12/0	1	32
Uno	ATmega328P	16 MHz	5 V/4–12 V	14/6	6/0	1	32
Leonardo	ATmega32U4	16 MHz	5 V/4–12 V	20/4	12/0	1	32
Yùn	ATmega32U4 AR9331 Linux	16 MHz 400 MHz	5 V	20/4	12/0	1	32
Ethernet	ATmega328P	16 MHz	5 V/4–12 V	14/4	6/0	—	32
Gemma	ATtiny85	8 MHz	3.3 V/ 4–16 V	3/2	1/0	—	8
MKRZero	SAMD21 Cortex-M0+ 32 bit low power ARM MCU	48 MHz	3.3 V	22/12	4 (ADC 8/10/ 12 bit)/1 (DAC 10 bit)	1	256

4.2 Arduino IDE

The Arduino integrated development environment (IDE) is an open-source software, and it makes it easy to write code and upload it to the board.

4.2.1 Steps to Install Arduino IDE

Step 1: Install Arduino IDE and open the window

To begin, install the Arduino IDE. Figure 4.4 shows the window of Arduino IDE.

Step 2: Choose the version of Arduino board

Arduino has many versions like UNO, MEGA, NANO, etc. Before starting the project, find out the suitable version by selecting the parameters according to the requirement. The most common board for beginners is the Arduino UNO. Choose the board and the serial port

sketch_jun17a | Arduino 1.6.5
File Edit Sketch Tools Help

```
void setup() {
  // put your setup code here, to run once:

}

void loop() {
  // put your main code here, to run repeatedly:

}
```

FIGURE 4.4
Window Arduino IDE.

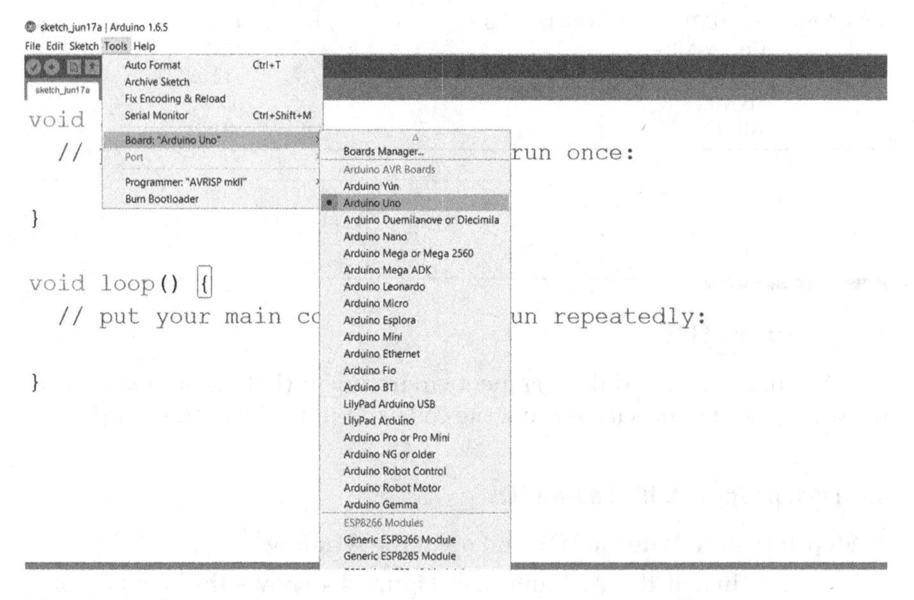

FIGURE 4.5
Selection of Arduino board.

in the Arduino IDE. To select the Arduino board, click on "Tool" and then click on "board." Figure 4.5 shows the selection of "Arduino Uno."

Step 3: Write and compile the program

Write the program in the Arduino IDE window. Then "RUN" the program. Figure 4.6 shows the window to compile the program.

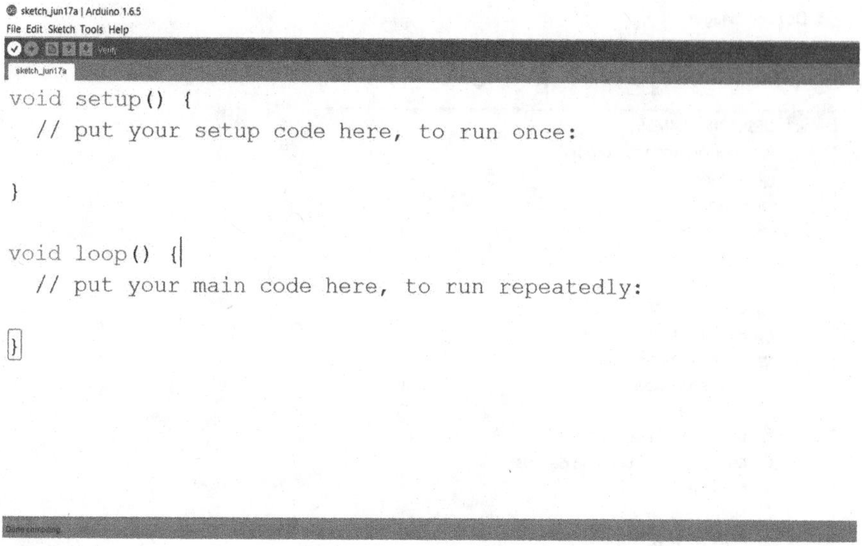

sketch_jun17a | Arduino 1.6.5
File Edit Sketch Tools Help

```
void setup() {
  // put your setup code here, to run once:

}

void loop() {
  // put your main code here, to run repeatedly:

}
```

Done compiling.

FIGURE 4.6
Compile the program.

Step 4: Connect Arduino with PC

Connect Arduino to the USB port of the PC with a USB cable. Every Arduino board has a different serial-port address (COM2, COM4, etc.), so it is required to reconfigure the port for each Arduino and select it in IDE. To check the port at which the Arduino is connected, right click on "PC" then select "manager"; a window will open. Then double click on "Device Manager." A window as shown in Figure 4.7 will open. Click on ports COM and LPT and port at which device is connected can be found.

Now click on the "Tool" head at the Arduino IDE window. Go to the port and select the same port number, which is found in the device manager (select COM1 or COM2, etc). Figure 4.8 shows the "COM38" as the serial port of the board.

Step 5: Upload program to Arduino board

Upload the program to the Arduino board. Figure 4.9 shows how to upload the program.

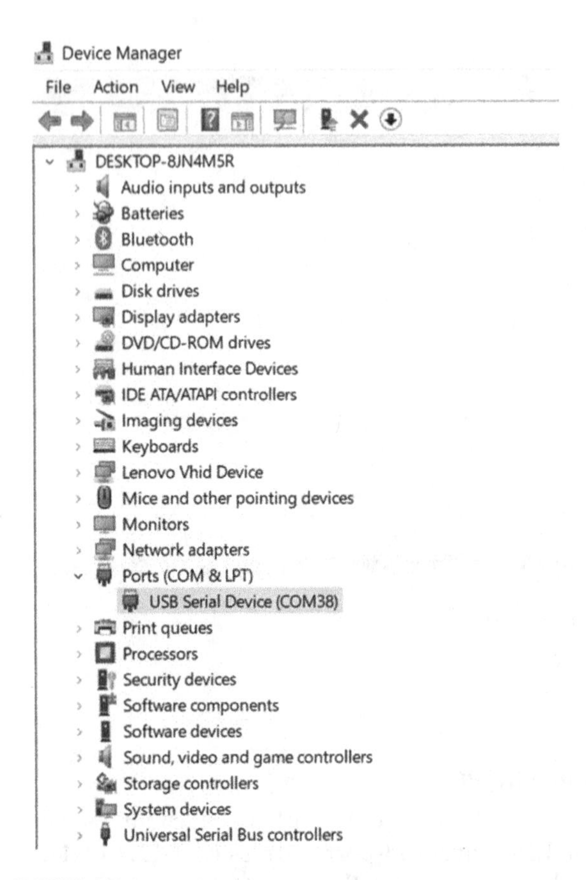

FIGURE 4.7
Window to check port of Arduino.

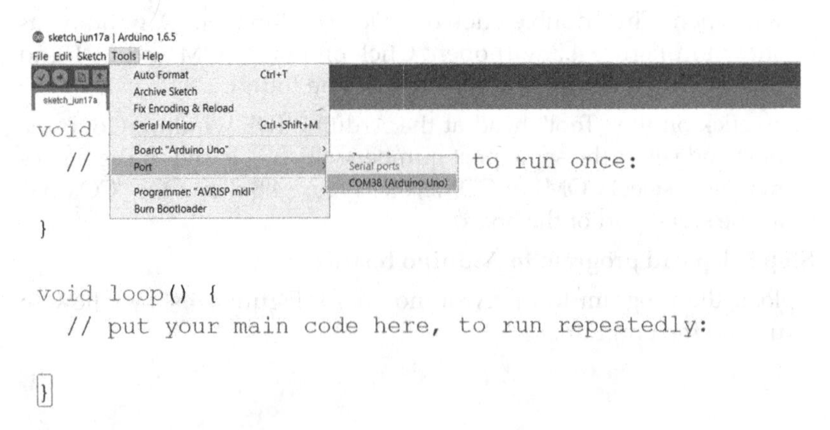

FIGURE 4.8
The serial port of the board.

sketch_jun17a | Arduino 1.6.5
File Edit Sketch Tools Help

```
void setup() {
  // put your setup code here, to run once:

}

void loop() {
  // put your main code here, to run repeatedly:

}
```

FIGURE 4.9
Window to upload the program.

4.3 Basic Commands for Arduino

1. **pinMode(x, OUTPUT);** // assigned pin number x as output pin where x is number of digital pin
2. **digitalWrite(x, HIGH);** // turn ON the pin number x as HIGH or ON where x is number of digital pin
3. **pinMode(x, INPUT);** // assigned pin number x as input pin where x is number of digital pin
4. **digitalRead(digital Pin);** // read the digital pin like 13 or 12 or 11 etc.
5. **analogRead(analog pin);** // read the analog pin like A0 or A1 or A2 etc.

4.4 LCD Commands

1. **lcd.begin(16, 2);** // initialize LCD 16*2 or 20*4
2. **lcd.print("RAJESH");** // print a string "RAJESH" on LCD
3. **lcd.setCursor(x, y);** // set the cursor of LCD at desired location where x is number of COLUMN and y
4. **lcd.print(LPU);** // print a LPU as integer on LCD
5. **lcd.Clear();** // clear the contents of LCD

4.5 Serial Communication Commands

1. **Serial.begin(baudrate);** // initialize serial communication to set baud rate to 600/1200/2400/4800/9600
2. **Serial.print("RAJESH");** // serial print fixed string with define baud rate on Tx line
3. **Serial.println("RAJESH");** // serial print fixed string with define baud rate and enter command on Tx line
4. **Serial.print("LPU");** // serial print int string with define baud rate on Tx line
5. **Serial.print("LPU");** // serial print int string with define baud rate and enter command on Tx line
6. **Serial.Write(BYTE);** // serial transfer the one byte on Tx line
7. **Serial.read();** // read one byte serial from Rx line

4.6 Play with LED and Arduino

A light-emitting diode (LED) is a device that can be used as an indicator. An LED has two terminals, anode and cathode. LEDs are available in a different colors. Figure 4.10 shows the LED.

FIGURE 4.10
Light-emitting diode.

Different colors can be used to represent different conditions. The color of the LED is due to the emission of light in specific regions of the visible light spectrum by different compounds.

To understand the working of LED, connect the anode of the LED to pin 4 of Arduino and the cathode to ground. Upload the sketch described in Section 4.4.1 to Arduino and observe the blinking of the LED.

Figure 4.11 shows the circuit diagram of the Arduino interfacing with the LED.

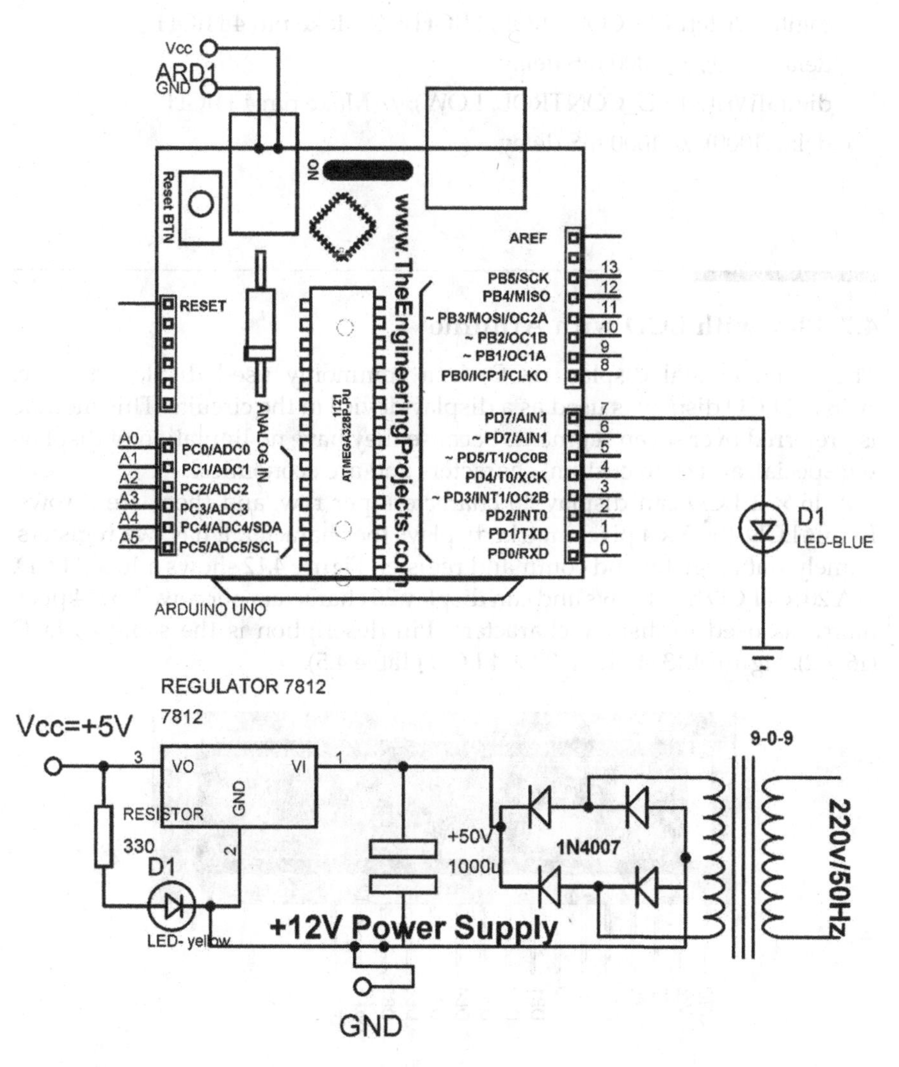

FIGURE 4.11
Circuit diagram to interface LED with Arduino.

4.6.1 Sketch

```
int LED_CONTROL=4;
void setup()
{
  pinMode(LED_CONTROL, OUTPUT); // initialize pin 4 as output pin
}
void loop()
{
  digitalWrite(LED_CONTROL, HIGH); // Make pin 4 HIGH
  delay(1000); // 1000 mS delay
  digitalWrite(LED_CONTROL, LOW); // Make pin 4 HIGH
  delay(1000); // 1000 mS delay
}
```

4.7 Play with LCD with Arduino

The liquid crystal display (LCD) is a commonly used display module. A 16 × 2 LCD display is used as a display device in the circuits. This module is preferred over seven segments because they have no limitation of displaying special, and even custom, characters and are economical.

A 16 × 2 LCD can display 16 characters per row, and there are 2 rows. In this LCD, the 5 × 4 pixel matrix displays the character. It has two registers, namely, data register and command register. Figure 4.12 shows a 16 × 2 LCD.

A 20 × 4 LCD has 4 rows and can display 20 characters per row. A 5 × 4 pixel matrix is used to display characters. Pin description is the same as LCD (16 × 2). Figure 4.13 shows a 20 × 4 LCD (Table 4.5).

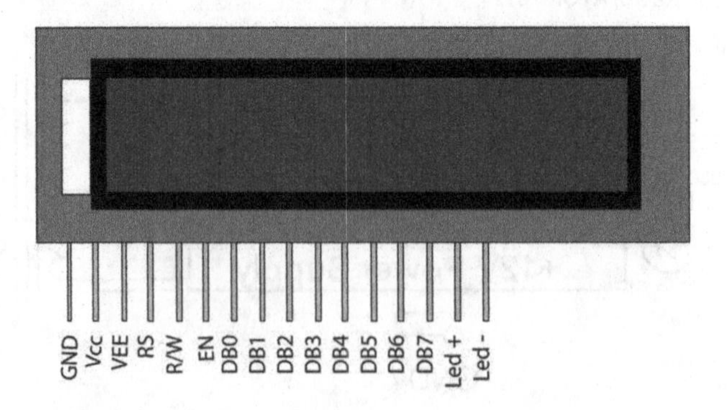

FIGURE 4.12
Liquid crystal display (16 × 2).

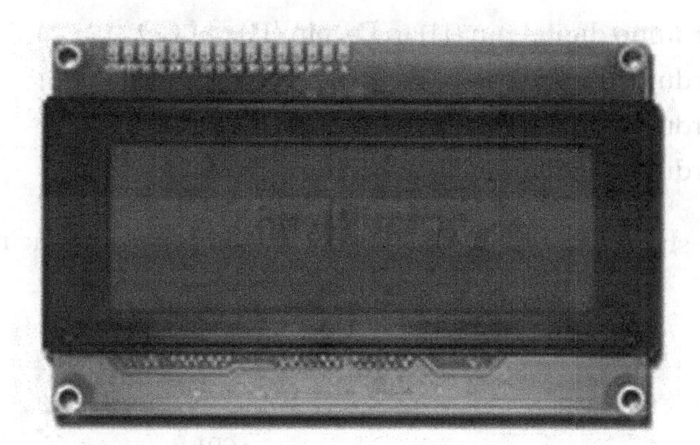

FIGURE 4.13
Liquid crystal display (20 × 4).

TABLE 4.5
LCD Pin Description

Pin	Description
Pin (1) Ground	Ground (0 V)
Pin (2) V_{CC}	Power supply (5 V)
Pin (3) V_{EE}	A variable resistor is used to adjust the contrast
Pin (4) Register Select	When low, it selects the command register, and if high, then it selects the data register
Pin (5) Read/Write	High to read the register and low to write on the register
Pin (6) Enable	Send data to data line when high to low pulse is given
Pin (7) DB0	
Pin (8) DB1	
Pin (9) DB2	8 bit data lines
Pin (10) DB3	
Pin (11) DB4	
Pin (12) DB5	
Pin (13) DB6	
Pin (14) DB7	
Pin (15) LED^+	Backlight Vcc (5 V)
Pin (16) LED^-	Backlight ground (0 V)

LCD connection

Connect the components as follows:

- Arduino digital pin (13) to RS pin (4) of LCD.
- Arduino digital pin (GND) to RW pin (5) of LCD.
- Arduino digital pin (12) to E pin (6) of LCD.

- Arduino digital pin (11) to D4 pin (11) of LCD.
- Arduino digital pin (10) to D5 pin (12) of LCD.
- Arduino digital pin (9) to D6 pin (13) of LCD.
- Arduino digital pin (8) to D7 pin (14) of LCD.

Figure 4.14 shows the circuit diagram of the Arduino interfacing with the LCD.

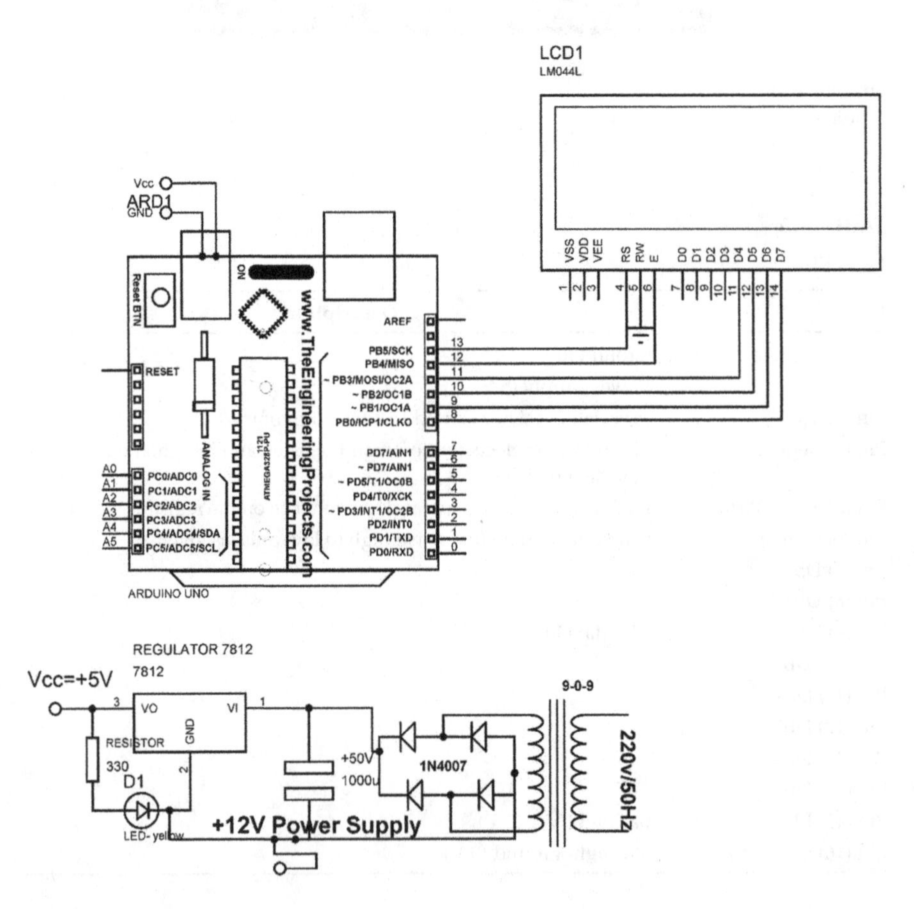

FIGURE 4.14
Circuit diagram to read LCD.

4.7.1 Sketch

```
#include <LiquidCrystal.h>
LiquidCrystal lcd(13, 12, 11, 10, 9, 8);
void setup()
{
  lcd.begin(20, 4); // Initialize LCD
  lcd.print("WELCOME"); // Print string on LCD
  delay(2000); // Delay 2000mS
  lcd.clear();
}
void loop()
{
  lcd.setCursor(0, 1); // set cursor of LCD
  lcd.print("ECE Department"); // Print string on LCD
  delay(2000); // Delay 2000mS
  lcd.setCursor(0, 2); // set cursor of LCD
  lcd.print("Rajesh Singh"); // Print string on LCD
  delay(2000); // Delay 2000mS
}
```

5

Basics of Raspberry Pi

5.1 Introduction to Raspberry Pi

Raspberry Pi is a low-cost mini-computer that can be connected to a computer monitor or TV. It is a device for exploring the computer computing with the help of a keyboard and mouse. A few optional parts can be connected with it for better accessibility. It can be programmed in languages like Scratch and Python. The processor frequency ranges from 700 MHz to 1.4 GHz for the Pi 3 model B+. The onboard memory ranges from 256 MB to 1 GB RAM. Raspberry Pi 3+ has a 1.4 GHz 64-bit quad core ARM cortex A53 processor. Figure 5.1a shows Raspberry Pi.

The first generation of Raspberry Pi 1 model "B" was launched in February 2012. An improved design Raspberry Pi 1 model "B+" was launched in 2014. A Raspberry Pi Zero has a smaller size and reduced input/output (I/O) with general-purpose input/output (GPIO) capabilities, and it was launched in November 2015. The Raspberry Pi Zero W, a version of the Zero with Wi-Fi and Bluetooth capabilities, was launched on February 28, 2014. The Raspberry Pi Zero WH was launched on January 12, 2018; it is a version of the Zero W with presoldered GPIO headers. Raspberry Pi 3 Model B was launched in February 2016 with an onboard Wi-Fi, Bluetooth 1.2 GHz, 64-bit quad core processor, and USB boot capabilities. The Raspberry Pi 3 Model B+ was launched in 2018 with a 1.4 GHz processor and a 300 Mbit/s by the internal USB 2.0 connection or 2.4/5 GHz dual-band Wi-Fi (100 Mbit/s). Figure 5.1b shows the pin diagram of Raspberry Pi 3 B+ model.

5.1.1 Raspberry Pi Components

1. **Power supply:** The Raspberry Pi can be powered with a micro-USB connection capable of supplying at least 400 mA at 5 V; standard mobile phone chargers are suitable for this purpose. If Raspberry Pi is powered up from a USB port of another computer, it may not supply the required current. Figure 5.2 shows the micro-USB power supply.

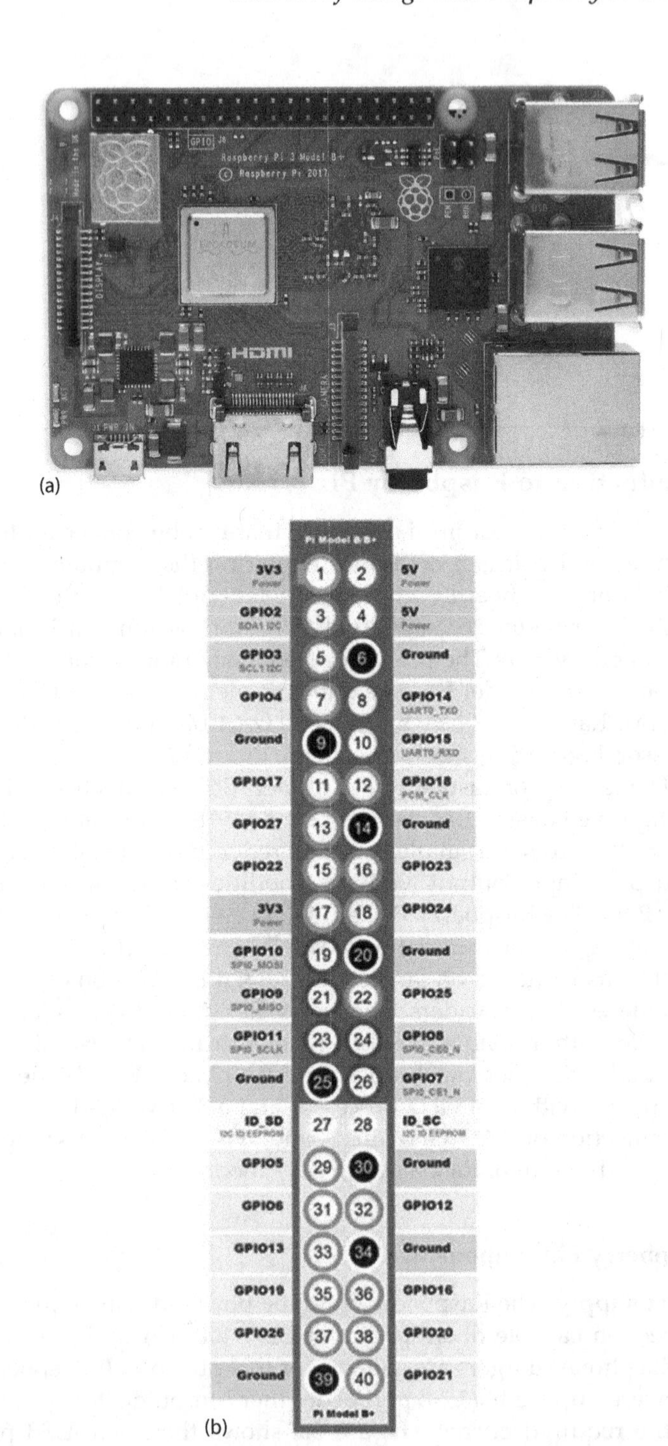

FIGURE 5.1
(a) Raspberry Pi. (b) Pin diagram of Raspberry Pi 3 B+ model.

FIGURE 5.2
Micro-USB power supply.

2. **Screen:** The screen is a monitor or television with HDMI or DVI input. It can also be connected with VGA with HDMI to VGA adapter.
3. **Keyboard and mouse:** Lower-power USB peripherals (keyboard and mouse) with less than 0.1 A power are suitable to connect with Raspberry Pi; more power-consuming devices may need separate power hubs.
4. **SD card:** The SD card used for Raspberry Pi should be at least Class 4, with minimum 8 GB of storage capability. Figure 5.3 shows the SD card. The SD card needs to be installed on an operating system from another computer. This installing process can be performed on Windows, Mac, and Linux. Operating systems like NOOBS and Raspbian are most commonly used operating systems to install on an SD card.

FIGURE 5.3
SD card.

5.2 Installation of NOOBS on SD Card

The primary requirement to install NOOBS is desktop, as it can't be installed on a laptop. Once it is installed, then it can be used on laptops.

Steps to install NOOBS on SD card

1. Format the SD card with the help of an SD card formatter. It can be downloaded from https://www.sdcard.org/downloads/formatter_4/.
2. Download the archive file for NOOBS from http://www.raspberrypi.org/downloads.
3. Extract the archive file of NOOBS and copy the contents of folder on the SD card.
4. Insert the SD card in the Raspberry Pi and power it up. A window will open when it boots, as shown in Figure 5.4, then click on Raspbian, and a new window will open as shown in Figure 5.5. Next, a Linux window for Raspberry Pi will open, and now it is ready to work.
5. Download the angry IP scanner software and get the IP address (Figure 5.6).
6. Now download a software "putty" or "mobaX term" and copy the IP in it that was found in the previous step and enter in the Linux terminal of Pi.
7. Enter the default user-id > pi and password > raspberry.

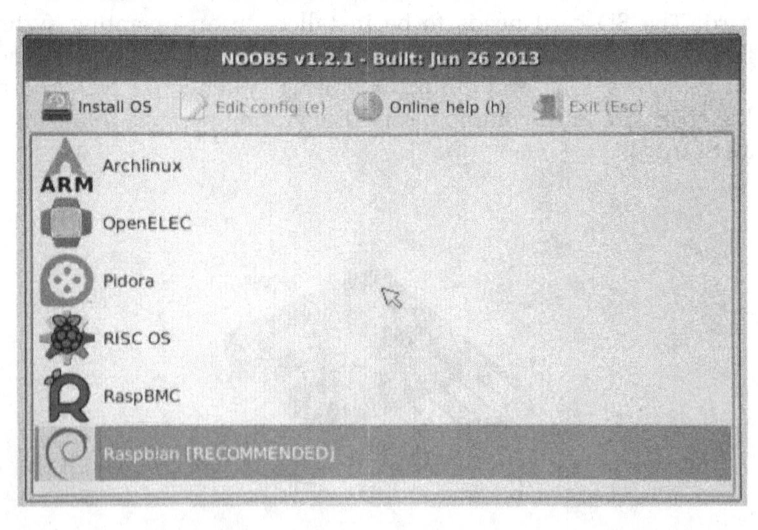

FIGURE 5.4
NOOBS window.

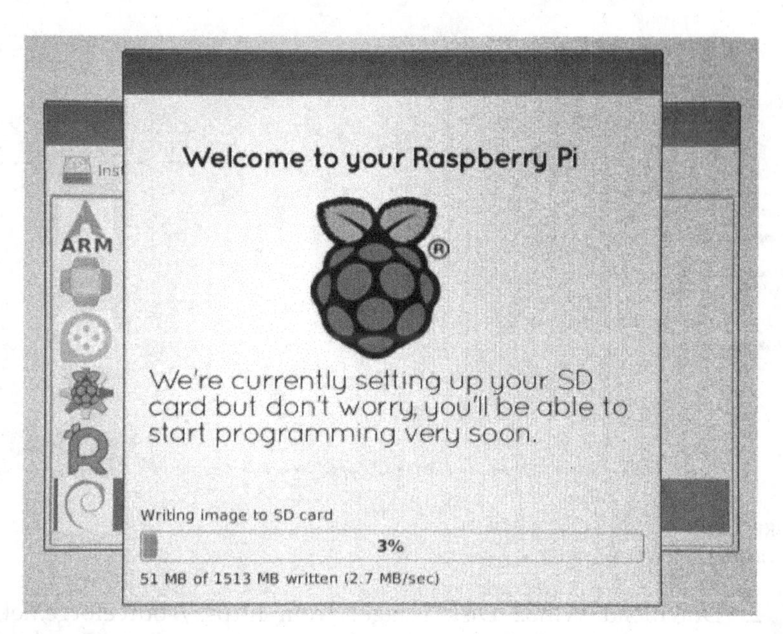

FIGURE 5.5
Raspberry Pi booting up.

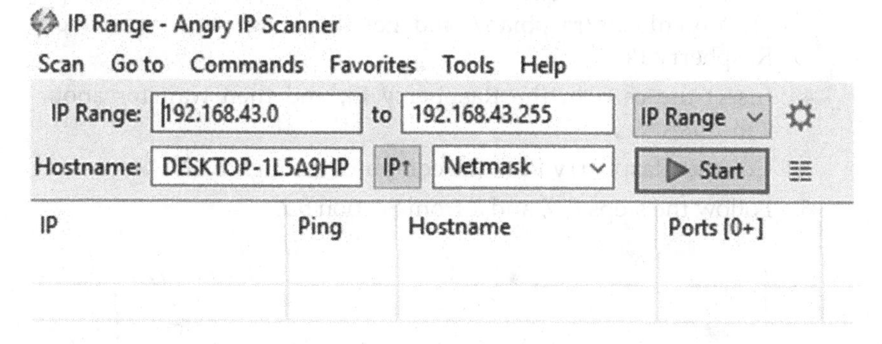

FIGURE 5.6
Angry IP scanner.

5.3 Installation of Raspbian on SD Card

Steps to install Raspbian on SD card

To complete the process, two softwares and one operating system (Raspbian) need to be downloaded.

1. Format the SD card with the help of an SD card formatter. It can be downloaded from https://www.sdcard.org/downloads/formatter_4/. Figure 5.7 shows the SD formatter window.

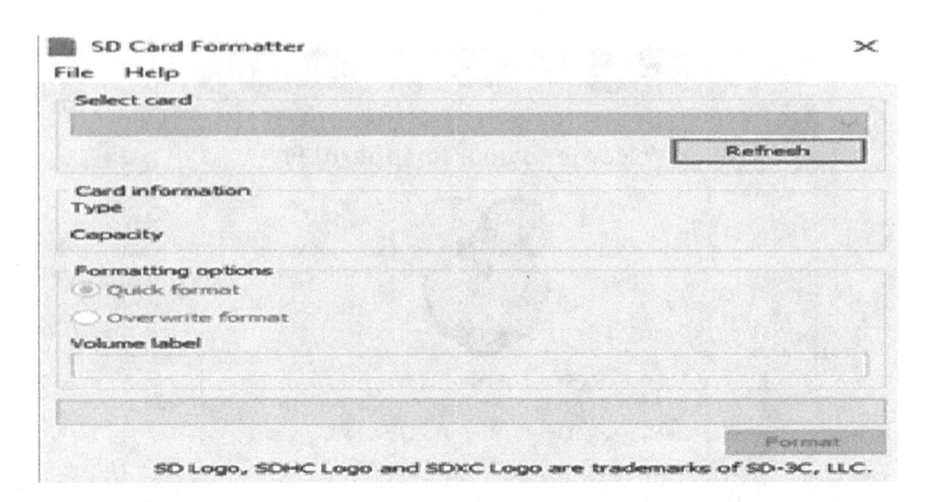

FIGURE 5.7
SD formatter.

2. Download Win32 Disk Imager from https://sourceforge.net/projects/win32diskimager/ for burning the image of Raspbian on the SD card. Figure 5.8 shows the Win32 disk imager window.
3. Download the Raspbian from https://www.raspberrypi.org/downloads/raspbian/ and get the image and burn it on Raspberry Pi.
4. Insert the card in the Raspberry Pi, and then wait for about 2 minutes.
5. Connect Raspberry Pi to the computer through LAN.
6. Follow the steps 6, 7, and 8 from Section 5.2.

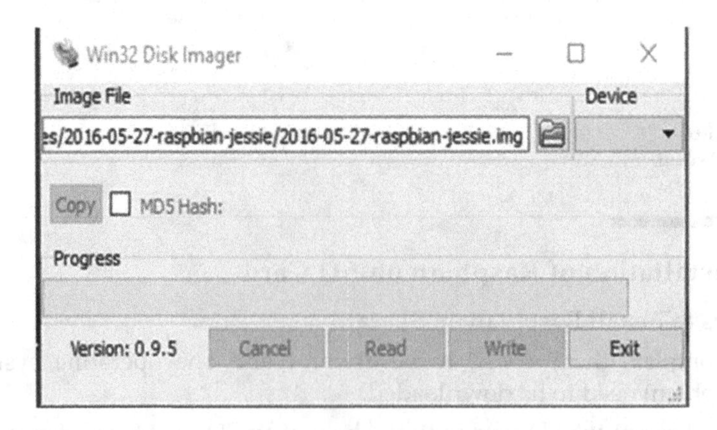

FIGURE 5.8
Win32 disk imager.

5.4 Terminal Commands

List of terminal commands:

1. *sudo apt-get update*: This command updates the package list.
2. *sudo apt-get upgrade*: This command downloads and installs the updated packages.
3. *sudo apt-get clean*: This command cleans an old package files.
4. *sudoraspi-config*: This command configurations the tool for Raspberry Pi.
5. *ls*: This command shows the list of directory contents.
6. *cd*: This command changes the directories.
7. *mkdir*: This command creates a directory.
8. *rmdir*: This command removes a directory.
9. *mv*: This command moves a file.
10. *tree -d*: This command shows a tree of directories.
11. *pwd*: This command shows the current directory.
12. *clear*: This command is for clearing the terminal window.
13. *sudo halt*: This command makes the Raspberry Pi shut down.
14. *sudo reboot*: This command is to restart the Raspberry Pi.
15. *startx:* This command is to start the desktop environment (LXDE).
16. *ifconfig:* This command is to find IP address of Raspberry Pi.
17. *nano:* This command is to edit a file.
18. *cat:* This command shows the content of a file.
19. *rm:* This command removes a file.
20. *cp:* This command is to copy a file or directory.
21. *locate:* This command is to locate a file.
22. *sudo:* This command is used in the Linux command line; sudo stands for "SuperUser Do."
23. *df:* This command is to see the available disk space in each of the partitions in the system.
24. *du:* This command shows the disk usage of a file in the system.
25. *tar:* "tar -cvf" is for creating a .tar archive, -xvf to untar a tar archive, -tvf to list the contents of the archive, etc.
26. *zip, unzip:* This command compresses the files into a zip archive, and unzip it to extract files.
27. *uname:* The command "uname -a" prints most of information about the system.

28. *chmod:* This command is to change the permissions of files or directories.
29. *ping:* This command is to check your connection to a server.
30. *sudo shutdown:* This command is to shut down the Raspberry Pi.

5.5 Installation of Libraries on Raspberry Pi

$ sudo apt-get install libblas-dev
$ sudo apt-get install liblapack-dev
$ sudo apt-get install gfortran
$ sudo apt-get install python-setuptools
$ sudoeasy_installscipy
$ ## previous might also work: python-scipy without all dependencies
$ sudo apt-get install python-matplotlib

5.6 Getting the Static IP Address of Raspberry Pi

Raspberry Pi gives a different IP address. Each time it gets "ON," this IP is known as current IP address. So each time, user needs to check for its IP. To avoid this problem, a static IP address needs to be created, which will not change in turning "OFF" the device.

Steps to create a static IP address:

1. Open the terminal window (Figure 5.9).
2. At the command line, write the command "ifconfig" (Figure 5.10).

   This command gives network-related information. The Figure 5.10 highlighted block shows the current IP address; here it is 192.168.0.16. Note this IP address, as this will be required later.
3. Open the configuration file.

   Now open the configuration file where a static IP address can be configured. The filename is *dhcpcd.conf,* and it is stored in the folder named *etc.* To make any change, use the nano editor in the terminal window with the help of the command: "sudonano / etc/dhcpcd.conf" (Figure 5.11). After this command, a window will open (Figure 5.12).

   In the screen, Figure 5.12, scroll down until you see: "# Example static IP configuration" (Figure 5.13).

FIGURE 5.9
Terminal window.

FIGURE 5.10
Command window for Raspberry Pi after command "ifconfig."

Now remove "#" from each line of this paragraph and change the last figures of the IP address from a range of 1 to 255 (Figure 5.14). Once changes are made, exit the window by using Ctrl+x and save the changes by click on "Yes" (Figure 5.15).

4. Reboot the Raspberry Pi by the command "reboot" (Figure 5.16).
5. Check the static IP address of Raspberry by the command "ifconfig" (Figure 5.17).

```
pi@raspberrypi                                                    -  □  ×
File  Edit  Tabs  Help
pi@raspberrypi:~ $ ifconfig
eth0: flags=4163<UP,BROADCAST,RUNNING,MULTICAST>  mtu 1500
        inet 192.168.0.16  netmask 255.255.255.0  broadcast 192.168.0.255
        inet6 2a02:2788:1014:63:f2bc:d1c7:c3e3:89e0  prefixlen 64  scopeid 0x0<global>
        inet6 fe80::894b:60fe:1b3:1e7a  prefixlen 64  scopeid 0x20<link>
        inet6 2a02:2788:1014:63:ba27:ebff:fef3:bec5  prefixlen 128  scopeid 0x0<global>
        ether b8:27:eb:f3:be:c5  txqueuelen 1000  (Ethernet)
        RX packets 345  bytes 39535 (38.6 KiB)
        RX errors 0  dropped 0  overruns 0  frame 0
        TX packets 121  bytes 10887 (10.4 KiB)
        TX errors 0  dropped 0 overruns 0  carrier 0  collisions 0

lo: flags=73<UP,LOOPBACK,RUNNING>  mtu 65536
        inet 127.0.0.1  netmask 255.0.0.0
        inet6 ::1  prefixlen 128  scopeid 0x10<host>
        loop  txqueuelen 1000  (Local Loopback)
        RX packets 9  bytes 524 (524.0 B)
        RX errors 0  dropped 0  overruns 0  frame 0
        TX packets 9  bytes 524 (524.0 B)
        TX errors 0  dropped 0 overruns 0  carrier 0  collisions 0

wlan0: flags=4163<UP,BROADCAST,RUNNING,MULTICAST>  mtu 1500
        inet 192.168.0.24  netmask 255.255.255.0  broadcast 192.168.0.255
        inet6 2a02:2788:1014:63:ba27:ebff:fea6:eb90  prefixlen 128  scopeid 0x0<global>
        inet6 fe80::1b6c:74b:31b3:9b99  prefixlen 64  scopeid 0x20<link>
        inet6 2a02:2788:1014:63:ab7:65ba:1ed5:7b74  prefixlen 64  scopeid 0x0<global>
        ether b8:27:eb:a6:eb:90  txqueuelen 1000  (Ethernet)
        RX packets 155  bytes 21187 (20.6 KiB)
        RX errors 0  dropped 0  overruns 0  frame 0
        TX packets 85  bytes 12838 (12.5 KiB)
        TX errors 0  dropped 0 overruns 0  carrier 0  collisions 0

pi@raspberrypi:~ $ sudo nano /etc/dhcpcd.conf
```

FIGURE 5.11
Nano editor for static IP address configuration.

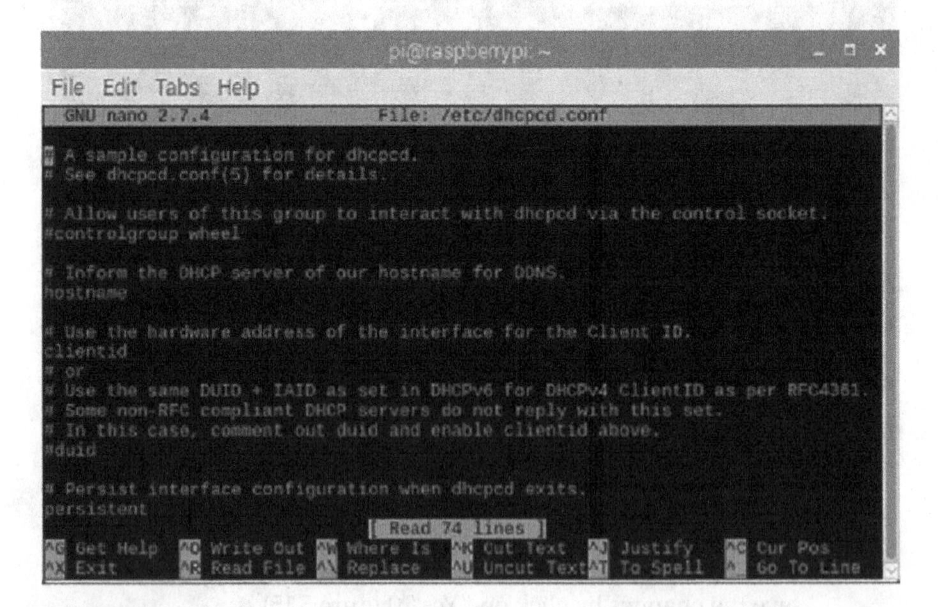

FIGURE 5.12
Window to configure IP address.

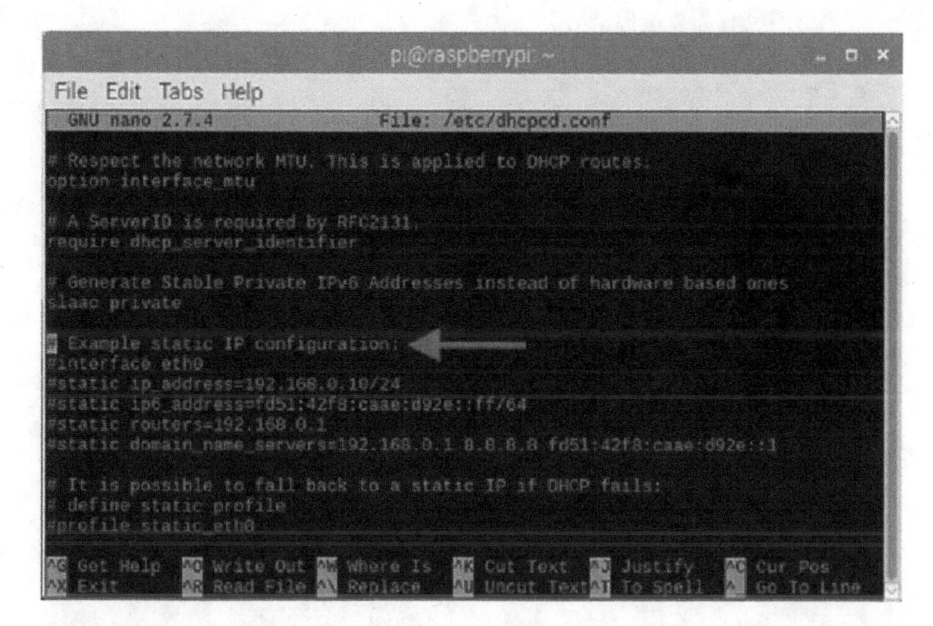

FIGURE 5.13
Static IP configuration.

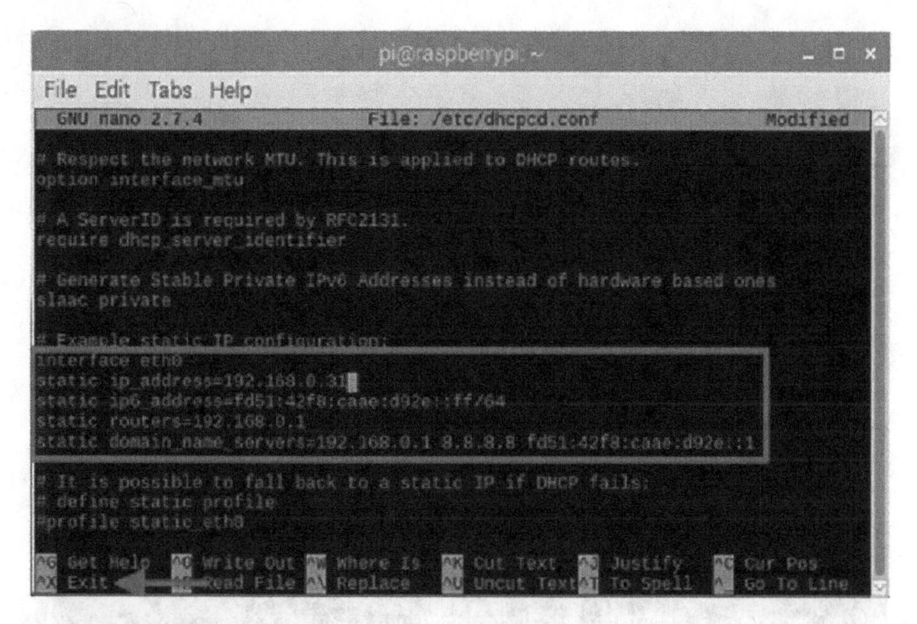

FIGURE 5.14
Static IP address.

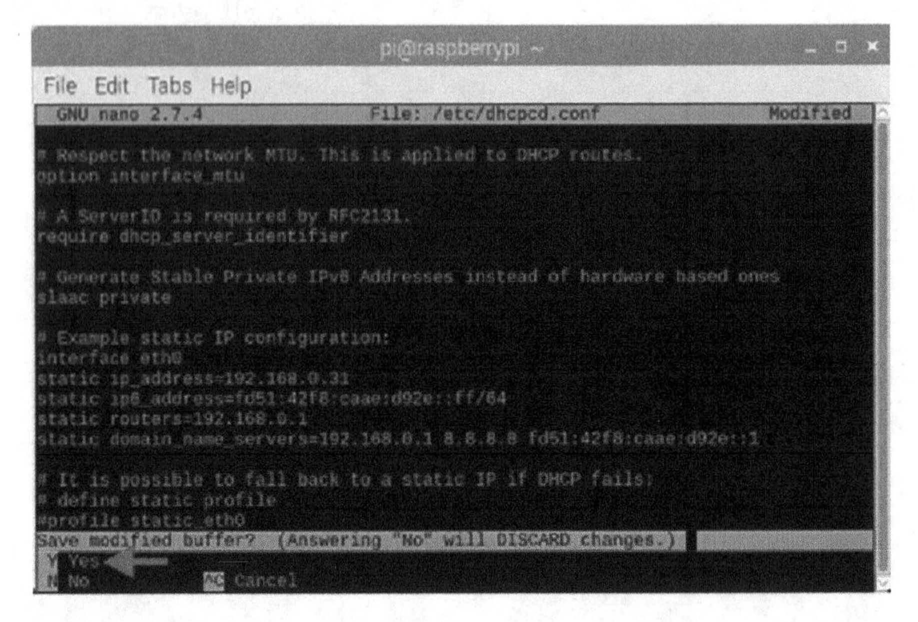

FIGURE 5.15
"Yes" to changes.

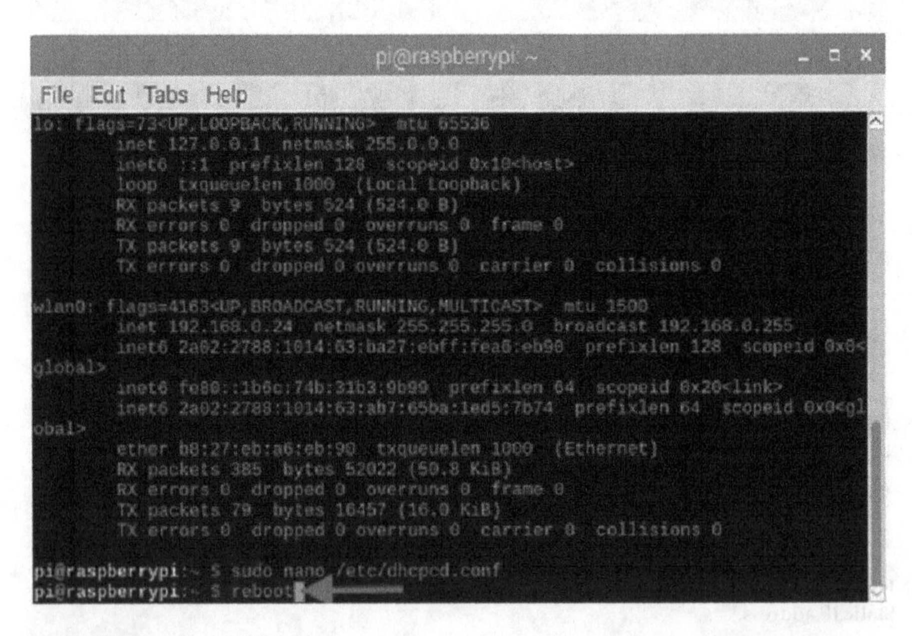

FIGURE 5.16
Reboot the Raspberry Pi.

FIGURE 5.17
Static IP address of Raspberry Pi.

5.7 Run a Program on Raspberry Pi

There are different ways to run a program on Raspberry Pi. A few of them are available to run a program and are as follows:

- rc.local
- .bashrc
- init.d tab
- systemd
- crontab

Here we will discuss "rc.local."

5.7.1 Editing rc.local

On Raspberry Pi, edit the file /etc/rc.local using the editor with root permissions: *"sudonano/etc/rc.local"*

Add commands to execute the Python program, save the file, and exit. If an infinite loop is running, then terminate the command by adding "&" at the last, e.g., *sudopython/home/pi/sample.py&*

The Pi runs this program at bootup before the start of other services. The Pi will not complete its boot process if the ampersand is not included and program runs continuously. The ampersand allows the command to run in a separate process and continue booting with running the main process. Now, reboot the Pi to test it: "sudo reboot."

5.8 Installing the Remote Desktop Server

The remote desktop server is helpful for a user to take control of a remote computer through a network connection. To control Raspberry Pi remotely, this server can be used. Open the Raspberry Pi terminal by using mouse and keyboard or by connecting via SSH (use raspi-config) to install the remote desktop server. On the Pi, install a package:

"sudo apt-get install xrdp"

Search for a "Remote Desktop Connection" for installing on a Windows PC (Figure 5.18).

By clicking on the option buttons, settings can be personalized, like resolution or keyboard/audio settings.

Enter the host name, or IP of the Raspberry Pi (or the DNS server) or the name of Pi (default: *raspberrypi*). If the PC is in the same network as the Pi, the login screen will appear. Find the IP address of Pi in the router's menu if it is unknown by opening 192.168.0.1 or 192.168.1.1. The internal IP of Pi will be in the format as follows: 192.168.x.xxx.

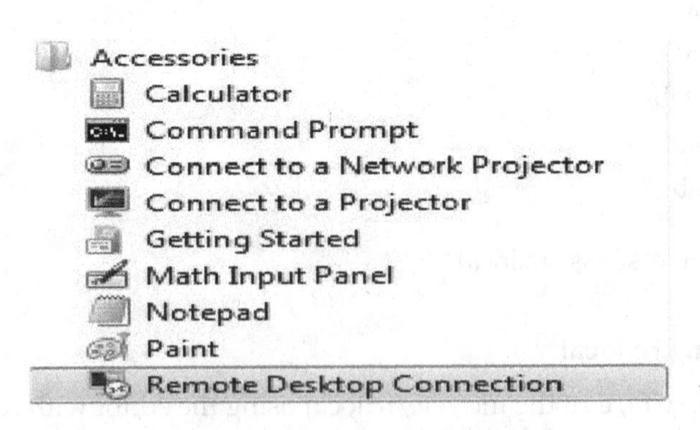

FIGURE 5.18
Remote desktop connection.

5.9 Pi Camera

Pi camera is a module with a Python interface. Install Picamera by the command *"sudo apt-get install python –picamera."*

5.9.1 Testing of the Camera

To test function of the camera, one image is taken and saved to /home/pi.

To capture an image, use command: *"raspistill -o /home/pi/image.jpg"*

When the command is executed, a camera preview will appear on the display. The preview will stay on for few seconds and then an image will be taken.

5.9.2 Raspberry Pi Camera as a USB Video Device

The Raspberry Pi camera is a standard USB video device (required by MJPG) in/dev. Load the "Video for Linux 2" (V4L2) module for the hardware (BCM2835) by using the command *"sudomodprobe bcm2835-v4l2."* A Video0 device file inside /dev directory will appear, after execution of the command. To verify, run the command

"ls /dev | grep vid" (Figure 5.19).

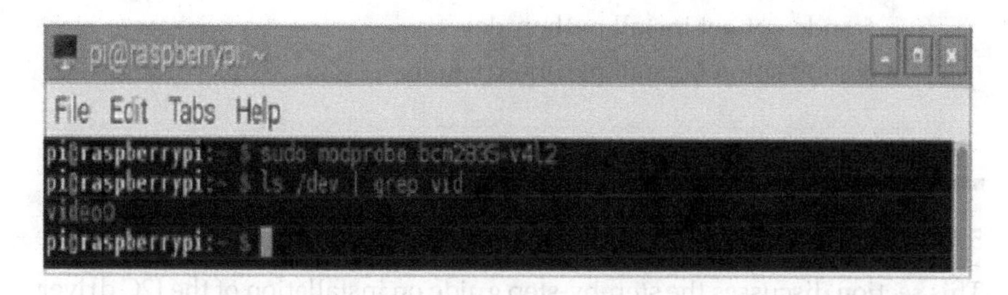

Verifying that USB video device file for Raspberry Pi camera is present inside /dev directory

FIGURE 5.19
Verify window.

5.10 Face Recognition Using Raspberry Pi

There are different methods for face detection like Haar cascades, HOG + Linear SVM, and CNNs.

Steps for face recognition

1. Configure the Picamera by using the command *"sudo apt-get install python –picamera."*
2. Update the system package list by using the command *"sudo apt-get update."*
3. Upgrade the installed packages to the latest versions with the help of the command *"sudo apt-get dist-upgrade."*
4. Install OpenCV if it is not done already. If RaspbianOS is installed on Raspberry, then use the following command to install openCV: *"sudo apt-get install python-opencv"* and *"sudo pip install imutils."*
5. Install Davis King's dlib toolkit software on the same Python virtual environment where OpenCV is installed into Raspberry Pi Face Recognition with the command "$pip install dlib."
6. Simply use pip to install Adam Geitgey's face_recognition module in Raspberry Pi Face Recognition by using the command "$pip install face_recognition."
7. Install my imutils package of convenience functions in Raspberry Pi Face Recognition with the command "$pip install imutils."
8. From a terminal window on Pi, type the following commands to fetch and install the RPiGPIO library:

   $ sudo apt-get install python-dev

   $ sudo apt-get install python-rpi.gpio

5.11 Installation of I2C Driver on Raspberry Pi

This section discusses the step-by-step guide on installation of the I2C driver for the Raspberry Pi.

1. Make a connection of Raspberry Pi with the internet.
2. The I2C driver is already installed in the new Raspbian distro, but by default it is disabled. To enable it, comment out a line by putting "#" in front, e.g., "sudonano /etc/modprobe.d/raspi-blacklist.conf" and then add a "#" on the third line (Figure 5.20).

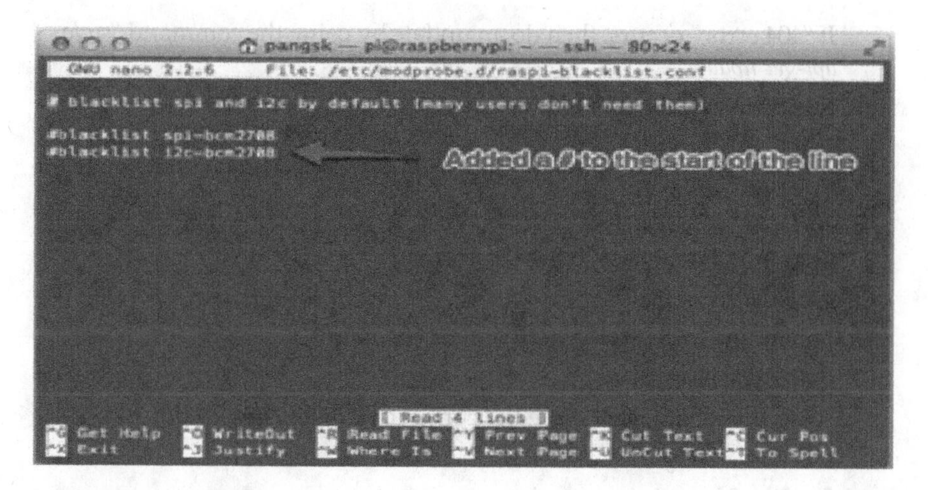

FIGURE 5.20
Window to enable I2C driver.

3. Press "ctrl+x" and then "y" to save and exit.
4. Edit the module's file by the command "sudonano /etc/modules" and add i2c-dev to a new line (Figure 5.21). Press "ctrl+x" then "y" to save and exit.
5. Install the i2c-tools package with the help of command *"sudo apt-get install i2c-tools."*

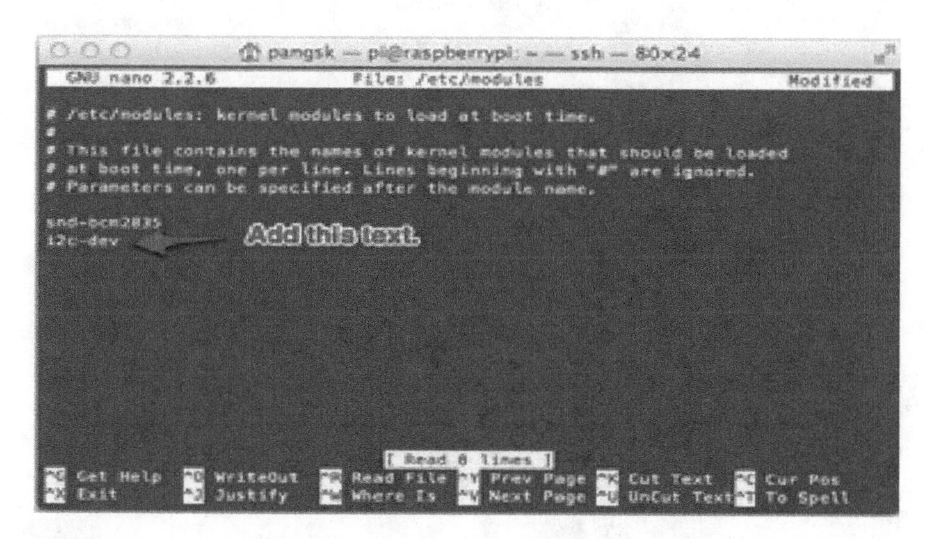

FIGURE 5.21
i2c-dev.

If "404 error" occurs, then do an update by using the command "*sudo apt-get update.*"

6. Run to install the i2c-tools again.

NOTE: The installation can take a few minutes, depending on how busy the server is.

7. Now add a new user to the i2c group:
   sudoadduser pi i2c
8. Reboot the PC with the help of the command:
   sudo shutdown -r now
9. After the reboot, test to see the connection of any device by:
   sudo i2cdetect -y 0
   For Rev 2 board command:
   sudo i2cdetect -y 1
   A window will appear (Figure 5.22).
10. Now install the python-smbuspython module with the help of command:
    sudo apt-get install python-smbus
11. Now Raspberry Pi is ready to use the i2c with Python.

```
pangsk — pi@raspberrypi: ~ — ssh — 80×24
pi@raspberrypi ~ $ i2cdetect -y 0
     0  1  2  3  4  5  6  7  8  9  a  b  c  d  e  f
00:          -- -- -- -- -- -- -- -- -- -- -- -- --
10: -- -- -- -- -- -- -- -- -- -- -- -- -- -- -- --
20: 20 -- -- -- -- -- -- -- -- -- -- -- -- -- -- --
30: -- -- -- -- -- -- -- -- -- -- -- -- -- -- -- --
40: -- -- -- -- -- -- -- -- -- -- -- -- -- -- -- --
50: -- -- -- -- -- -- -- -- -- -- -- -- -- -- -- --
60: -- -- -- -- -- -- -- -- -- -- -- -- -- -- -- --
70: -- -- -- -- -- -- -- --
pi@raspberrypi ~ $
```

FIGURE 5.22
Window after reboot test command.

5.12 Serial Peripheral Interface with Raspberry Pi

Adafruit Occidentalis 0.2 or later is preconfigured with serial peripheral interface (SPI) support. For Raspbian, few configuration changes are required.

The I2C driver is pre-installed in new Raspbian distro, but it is disabled by default. To enable it, comment out a line by putting "#" in front, e.g., "sudonano /etc/modprobe.d/raspi-blacklist.conf" and then add a "#" to the third line (Figure 5.20).

1. Edit the file */etc/modules* using the command "sudonano*/etc/modprobe.d/raspi-blacklist.conf*" and add a "#" to "blacklist spi-bcm2408," which will become "#blacklist spi-bcm2408."
2. The SPI feature of the Raspberry Pi is supported by a Python library, which carries out SPI communication with a Python program. To install this, install Git and then issue the following commands:

   $ cd ~

   $ sudo apt-get install python-dev

   $ git clone git://github.com/doceme/py-spidev

   $ cdpy-spidev/

   $ sudo python setup.py install
3. Reboot the Pi, and it is ready for SPI.

5.13 Programming a Raspberry Pi

To program a Raspberry Pi, the pin mode can be defined in two methods, BCM and BOARD. BCM stands for Broadcom SOC channel, and BCM refers to the GPIO number (i.e., GPIO21, GPIO22, etc.). BOARD refers to the pin number of the Pi board.

Commands:

importRPi.GPIO as X

This is to access GPIO pins of Raspberry Pi as the name "X" indicated in the command.

import time

Imports the Time library so that we can pause the script later on.

X.setmode(X.BCM)

Pins are referred as GPIO numbers, as BCM mode is selected.

X.setwarnings(False)

This command is not to print GPIO warning messages on the screen.

X.setup(21,X.OUT)
This is to define GPIO 21 as output pin.
print("Two LED ON")
This is to print the string on the terminal.
X.output(21, True) or X.output(21, HIGH)
This is to turn the GPIO 21 pin "on."
time.sleep(1)
To pause the Python program for 1 s.
exceptKeyboardInterrupt:
End program with keyboard.
GPIO.cleanup()
Reset GPIO settings.

5.14 Play with LED and Raspberry Pi

The interfacing of LED with Raspberry Pi is the simplest way to understand the basic command. The circuit is comprised of two LEDs, two 330-ohm resistors, and jumper wires. To understand the working of LEDs, connect the anode of LED1 to GPIO21 and LED2 to GPIO22 and cathode to the ground. Write the recipe described in Section 5.14.1 and observe the blinking of the LEDs.

Figure 5.23 shows the circuit diagram of the Raspberry Pi interfacing with LED.

5.14.1 Recipe for LED Blink

```
importRPi.GPIO as RAJ
import time as wait
LED_PIN1=20 # assign LED_PIN1 to pin 20
LED_PIN2=21 # assign LED_PIN1 to pin 21
RAJ.setmode(RAJ.BCM) # use pi pins in BCM mode
RAJ.setwarnings(False) # remove the warnings
RAJ.setup(LED_PIN1,RAJ.OUT) # make pin 21 to output
RAJ.setup(LED_PIN2,RAJ.OUT) # make pin 20 to output
try:
while True:
RAJ.output(LED_PIN1, True) # make pin 20 to HIGH
RAJ.output(LED_PIN2, True) # make pin 21 to HIGH
```

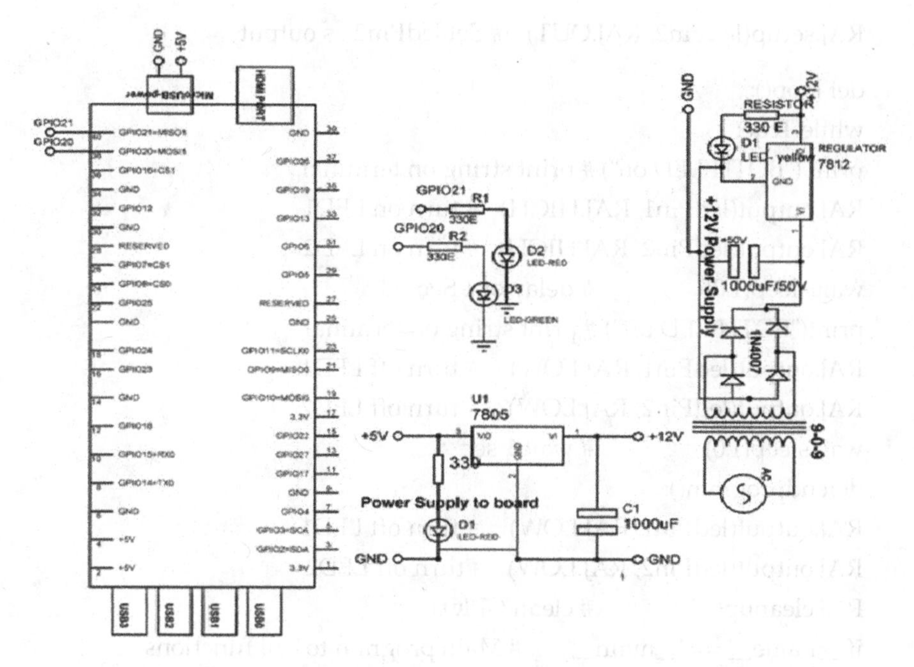

FIGURE 5.23
Circuit diagram of the Raspberry Pi interfacing with LED.

```
print("Two LED ON") # print string on terminal
wait.sleep(1)
RAJ.output(LED_PIN1, False) # make pin 20 to LOW
RAJ.output(LED_PIN2, False) # make pin 20 to LOW
print("Two LED OFF") # print string on terminal
wait.sleep(1)
exceptkeyboardInterrupt:
RAJ.cleanup()
```

5.14.1.1 Recipe for LED Blink Using Function

```
importRPi.GPIO as RAJ # RPi.GPIO can be referred as GPIO
import time as wait # import time library
ledPin1 = 21  # connect led1 on 21 pin
ledPin2 = 20  # connect led2 on 20 pin
def setup():
RAJ.setmode(RAJ.BOARD)        # GPIO Numbering of Pins
RAJ.setup(ledPin1, RAJ.OUT)   # Set ledPin1 as output
```

```
RAJ.setup(ledPin2, RAJ.OUT)  # Set ledPin2 as output

def loop():
while True:
print ("BOTH LED on") # print string on terminal
RAJ.output(ledPin1, RAJ.HIGH)  # turn on LED1
RAJ.output(ledPin2, RAJ.HIGH)  # turn on LED2
wait.sleep(1.0)              # delay of 1 Sec
print("BOTH LED off") # print string on terminal
RAJ.output(ledPin1, RAJ.LOW)   # turn off LED1
RAJ.output(ledPin2, RAJ.LOW)   # turn off LED2
wait.sleep(1.0)              # wait 1 sec
defendprogram():
RAJ.output(ledPin1, RAJ.LOW)   # turn off LED1
RAJ.output(ledPin2, RAJ.LOW)   #turn off LED2
RAJ.cleanup()                # clean GPIOs
if __name__ == '__main__':    # Main program to call functions
setup()
try:
loop()
exceptKeyboardInterrupt:  # keyboard interrupt to stop program
endprogram()
```

5.15 Reading the Digital Input

The digital inputs at Raspberry Pi can be read by two methods named as pull down and pull up, by setting the GPIO pins as active LOW or active HIGH output.

```
importRPi.GPIO as RAJ
RAJ.setup(23, RAJ.IN, pull_up_down=RAJ.PUD_DOWN)
RAJ.setup(24, RAJ.IN, pull_up_down=RAJ.PUD_UP)
```

These commands enable a pull-down resistor on pin 23 and a pull-up resistor on pin 24. The Pi is looking for a high voltage on pin 23 and a low voltage on pin 24. These are required to define in the loop, so that these can constantly check the pin voltage. To understand the concept, consider a small program for switch.

5.15.1 Recipe

```
importRPi.GPIO as RAJ
RAJ.setmode(GPIO.BCM) # use pi in BCM mode
RAJ.setup(23, RAJ.IN, pull_up_down = RAJ.PUD_DOWN) # set pin as
   input
RAJ.setup(24, RAJ.IN, pull_up_down = RAJ.PUD_UP) # set pin as input
while True:
if(RAJ.input(23) ==1):
print("pressed button 1") # print string on terminal
if(RAJ.input(24) == 0):
print("pressed button2") # print string on terminal
RAJ.cleanup() # clean all GPIOs
```

5.16 Reading an Edge-Triggered Input

The program in Section 5.15 is to detect whether either button is pressed or not. If the button is to trigger an action or command only one time, for this the GPIO library has a built-in function for rising-edge and falling-edge. A rising-edge is defined by the instant when signal changes from low to high, but it can only detect the change. Similarly, the falling-edge is the instant when the signal is from high to low.

The circuit of interfacing of the switch with Raspberry Pi is comprised of an LED, a resistor, a switch, and jumper wires. To understand the working of the switch, connect the components as shown in Figure 5.24. Write the recipe described in Sections 5.16.1 through 5.16.3 to observe the working of the switch in different configurations.

5.16.1 Reading Switch in Pull-Down Configuration

Connections:

- Connect one terminal of the switch to ground and other terminal to GPIO21 of Raspberry Pi.
- Connect one resistor of 10 K between +5 V and GPIO21 of Raspberry Pi.
- Connect the anode of the LED to GPIO20 of Raspberry Pi through a resistor and the cathode to ground.

Figure 5.24 shows the circuit diagram of the Raspberry Pi interfacing with the switch.

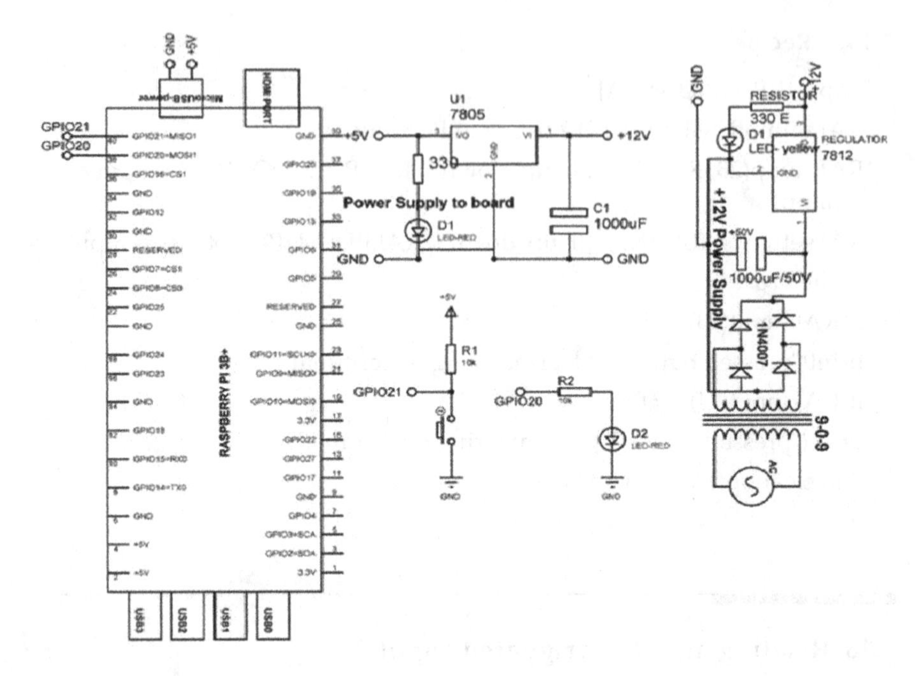

FIGURE 5.24
Circuit diagram of the Raspberry Pi interfacing with switch in "Pull-down" configuration.

5.16.1.1 Recipe for Pull-Down Configuration

```
importRPi.GPIO as RAJ
import time as wait # add time lib
button_pin=21# assign button_pin name to GPIO21
led_pin=20 # assign led_pin name to GPIO20
RAJ.setmode(GPIO.BCM) # use Pi pins as BCM
RAJ.setup(button_pin, RAJ.IN, pull_up_down = RAJ.PUD_UP) # GPIO
    as input
RAJ.setup(led_pin, RAJ.OUT) # make led_pin as output
try:
while True:
button_var = RAJ.input(button_pin)# read pin 21
ifbutton_var== False:
RAJ.output(led_pin, True) # make led_pin to HIGH
print("Switch Pressed..") # print string on terminal
wait.sleep(0.2) # delay of 200mSec
else:
```

```
RAJ.output(led_pin, False) # make led_pin to LOW
print("switch not pressed...") # print string on terminal
wait.sleep(0.2) # delay of 200mSec
exceptkeyboardInterrupt:
print("keyboard interrupt detected")
RAJ.cleanup() // clean all GPIOs
```

5.16.2 Reading Switch in Pull-Up Configuration

Connections:

- Connect one terminal of the switch to +5 V and the other terminal to GPIO21 of Raspberry Pi.
- Connect one resistor of 10 K between the ground and GPIO21 of Raspberry Pi.
- Connect the anode of the LED to GPIO20 of Raspberry Pi through a resistor and the cathode to ground.

Figure 5.25 shows the circuit diagram of the Raspberry Pi interfacing with the switch.

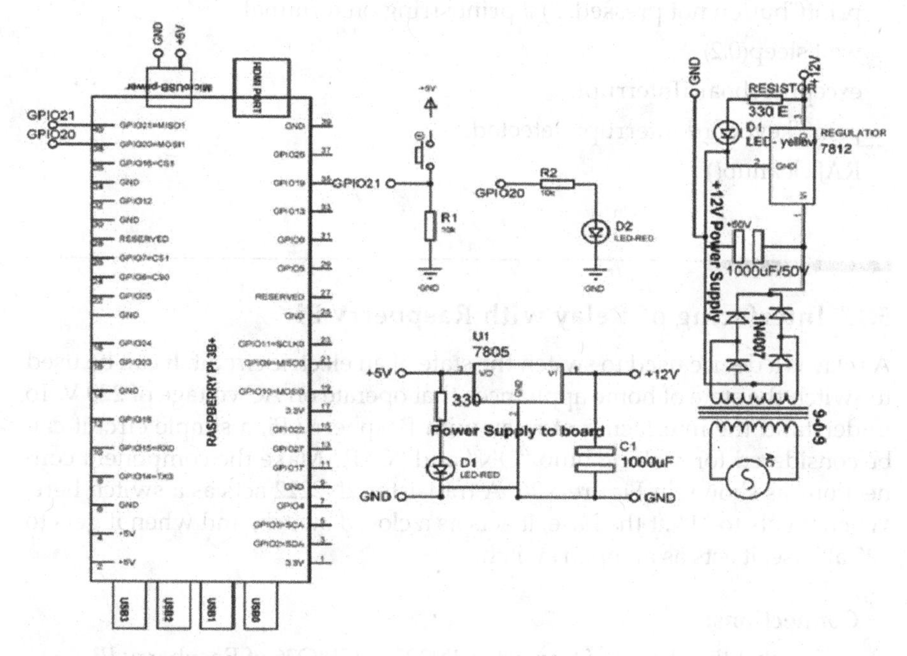

FIGURE 5.25
Circuit diagram of the Raspberry Pi interfacing with switch in "Pull-up" configuration.

5.16.2.1 Recipe for Pull-Up Configuration

```
importRPi.GPIO as RAJ
import time as wait
button_pin=21 # assign name to pin 21
led_pin=23 # assign name to pin 23
RAJ.setmode(RAJ.BCM) // use pins of pi in BCM mode
RAJ.setup(button_pin, RAJ.IN, pull_up_down = RAJ.PUD_DOWN)
   # set pin 21 as input
RAJ.setup(led_pin,RAJ.OUT) # set pin 23 as output
try:
while True:
button_VAR = RAJ.input(button_pin) # read switch
ifbutton_VAR == True:
RAJ.output(led_pin, True) # make pin 23 to HIGH
print("button pressed...") # print string on terminal
wait.sleep(0.2)
else:
RAJ.output(led_pin, False) # make pin 23 to LOW
print("button not pressed...") # print string on terminal
wait.sleep(0.2)
exceptkeyboardInterrupt:
print 'keyboard interrupt detected'
RAJ.cleanup()
```

5.17 Interfacing of Relay with Raspberry Pi

A relay is a device used to switch the state of an electric circuit. It can be used to switch the state of home appliances that operate on AC voltage of 220 V. To understand the interfacing of relay with Raspberry Pi, a simple circuit can be considered for making bulb "ON" and "OFF." Make the component connections as shown in Figure 5.26. A transistor 2N2222 acts as a switch here. When it gets to "1" at the base, it acts as a closed switch, and when it gets to "0" at base, it acts as an open switch.

Connections:

- Connect the "base" of transistor 2N222 to GPIO26 of Raspberry Pi.
- Connect "emitter" of transistor to the ground.

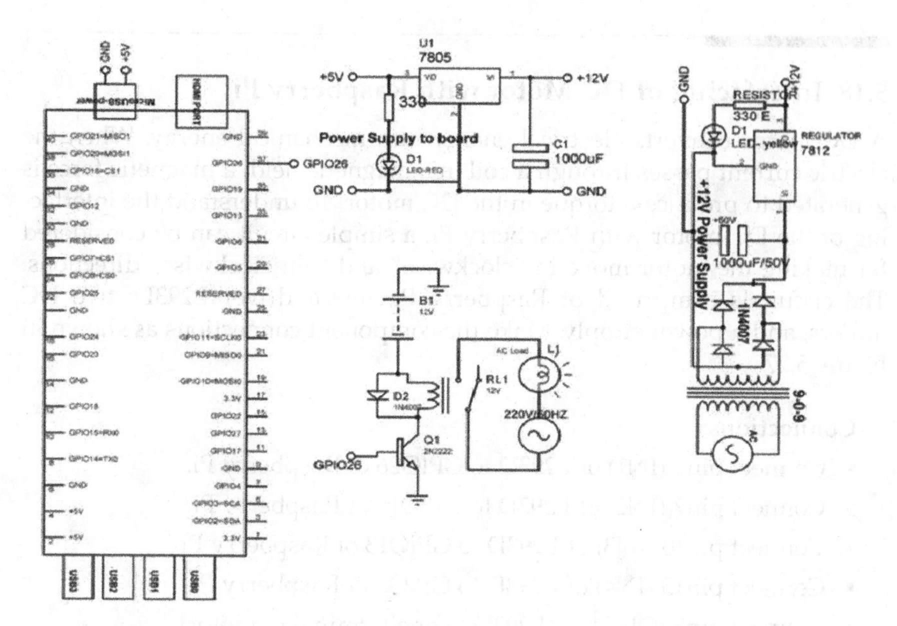

FIGURE 5.26
Circuit diagram for interfacing of relay with Raspberry Pi.

- Connect the "collector" of transistor to "L2" of relay.
- Connect the positive terminal of +12 V battery to "L1" of relay.
- Connect a diode 1N4004 across "L1" and "L2."
- Connect one terminal of AC to "common" of the relay and other one to the terminal of the AC load (bulb).
- Connect the other terminal of AC load to the "NO" terminal of the relay.

5.17.1 Recipe

```
importRPi.GPIO as MAHE # import pi GPIOs lib
import time as wait # wait time lib
MAHE.setmode(MAHE.BCM) # use pi pins in BCM mode
MAHE.setup(26, MAHE.OUT) # set pin 26 to output pin
while (True):
GPIO.output(26, MAHE.HIGH) // make pin 26 pin to HIGH
print(' relay ON') # print string on terminal
wait.sleep(5) # delay of 5 Sec
GPIO.output(26, GPIO.LOW)
print('relay OFF') # print string on terminal
wait.sleep(5) # delay of 5 Sec
```

5.18 Interfacing of DC Motor with Raspberry Pi

A DC motor converts electrical energy into mechanical energy. When the electric current passes through a coil in a magnetic field, a magnetic force is generated to produce a torque in the DC motor. To understand the interfacing of the DC motor with Raspberry Pi, a simple circuit can be considered for making the motor move in "clockwise" and "anticlockwise" directions. The circuit is comprised of Raspberry Pi, motor driver L293D, two DC motors, and a power supply. Make the component connections as shown in Figure 5.27.

Connections:

- Connect pin2 (IN1) of L293D to GPIO26 of Raspberry Pi.
- Connect pin7 (IN2) of L293D to GPIO19 of Raspberry Pi.
- Connect pin10 (IN3) of L293D to GPIO13 of Raspberry Pi.
- Connect pin15 (IN4) of L293D to GPIO6 of Raspberry Pi.
- Connect pin3 (OUT1) of L293D to one terminal of motor1.
- Connect pin6 (OUT2) of L293D to another terminal of motor1.

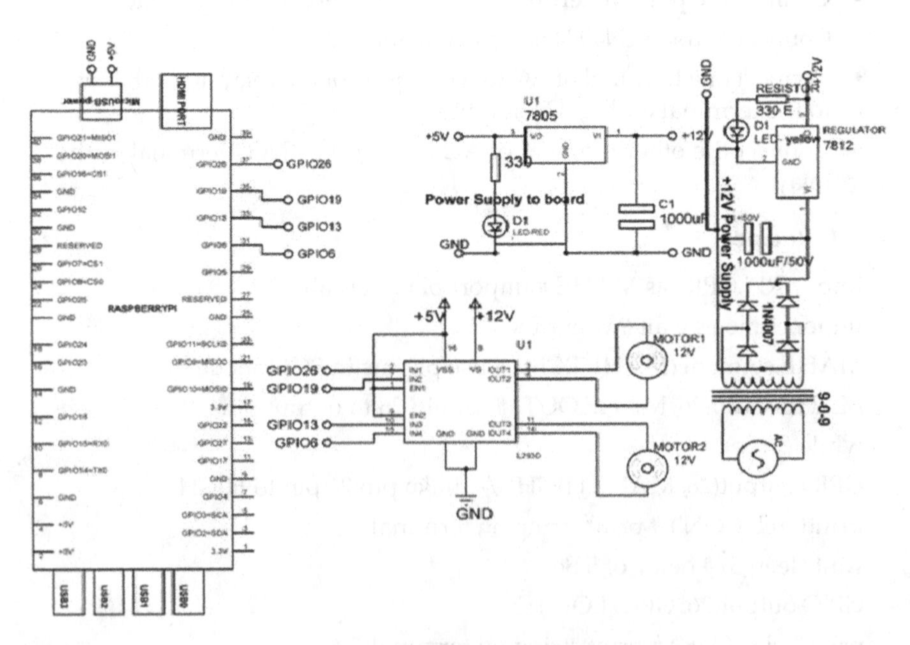

FIGURE 5.27
Circuit diagram for interfacing of DC motors with Raspberry Pi.

- Connect pin11 (OUT3) of L293D to one terminal of motor2.
- Connect pin14 (OUT4) of L293D to another terminal of motor2.
- Connect pin8 of L293D to +12 V DC.
- Connect pin1, 9, and 16 of L293D to +5 V DC.
- Connect "GND" pins of L293D to ground.

5.18.1 Recipe

```
importRPi.GPIO as MAHI #import GPIO as MAHI
import time as wait #include time library
M1_PIN1=26 # assign variable to GPIO26
M1_PIN2=19 # assign variable to GPIO19
M2_PIN1=13 # assign variable to GPIO13
M2_PIN2=6 # assign variable to GPIO6

MAHI.setmode(MAHI.BCM) # set pi as BCM
MAHI.setup(M1_PIN1,MAHI.OUT) #setting GPIO26 as OUTPUT
MAHI.setup(M1_PIN2,MAHI.OUT) #setting GPIO19 as OUTPUT
MAHI.setup(M2_PIN1,MAHI.OUT) #setting GPIO13 as OUTPUT
MAHI.setup(M2_PIN2,MAHI.OUT) #setting GPIO6 as OUTPUT

try:
while True:
MAHI.output(M1_PIN1, True) #make pin 26 to HIGH
MAHI.output(M1_PIN2, False) #make pin 19 to LOW
MAHI.output(M2_PIN1, True) #make pin 13 to HIGH
MAHI.output(M2_PIN2, False) #make pin 6 to LOW
print("Both Motor Forward") #print string on terminal
wait.sleep(10) # delay of 10 sec
MAHI.output(M1_PIN1, False) #make pin 26 to LOW
MAHI.output(M1_PIN2, True) #make pin 19 to HIGH
MAHI.output(M2_PIN1, False) #make pin 13 to LOW
MAHI.output(M2_PIN2, True) #make pin 6 to HIGH
print("Both Motor Reverse") #print string on terminal
wait.sleep(10) # delay of 10 sec
exceptkeyboardInterrupt:
MAHI.cleanup() # clean all GPIOs
```

5.18.2 Recipe to Control One Motor Using Function

```
importRPi.GPIO as MAHI
import time as wait
# Pins for Motor Driver Inputs
M11 = 26 #assign variable to GPIO26
M12= 19 #assign variable to GPIO19
def setup():
MAHI.setmode(MAHI.BCM) # choose BCM mode
MAHI.setup(M11,MAHI.OUT) # make M11 as output
MAHI.setup(M12,MAHI.OUT) # make M12 as output
MAHI.setup(M1E,MAHI.OUT) # make M1E as output

def loop():
        # forward movement
MAHI.output(M11,MAHI.HIGH) # set M11 to HIGH
GPIO.output(M12,MAHI.LOW) # set M12 to LOW
GPIO.output(M1E,MAHI.HIGH) # set M1E to HIGH
wait.sleep(5) // delay of 5 Sec
        # for reverse movement
MAHI.output(M11,MAHI.LOW) # set M11 to LOW
MAHI.output(M12,MAHI.HIGH) # set M12 to HIGH
MAHI.output(M1E,MAHI.HIGH) # set M1E to HIGH
wait.sleep(5) // delay of 5 Sec
        # for stop movement
MAHI.output(M1E,MAHI.LOW) # set M1E to LOW

def destroy():
MAHI.cleanup() # clean GPIOs
if __name__ == '__main__': # Program start from here
setup()
try:
loop()
exceptKeyboardInterrupt:
destroy() # stop the execution
```

5.19 Interfacing of LCD with Raspberry Pi

A liquid-crystal display is a display device. It does not emit light; instead, it has a blacklight to produce the image. It can be used for displaying the message or sensory data in a system. The circuit is comprised of Raspberry Pi, LCD, and a power supply. Make the component connections as shown in Figure 5.28.

Connections:

- Connect pin1 (Vss), pin5, (RW) and pin16 (LED-) of LCD to ground.
- Connect pin2 (Vcc) and pin15 (LED+) of LCD to +5 V DC.
- Connect pin3 (V_{EE}) of LCD to output of 10 K potentiometer and connect other two terminals of potentiometer to +5 V DC and ground.
- Connect pin4 (RS) of LCD to GPIO26 of Raspberry Pi.
- Connect pin6 (E) of LCD to GPIO19 of Raspberry Pi.
- Connect pin11 (D4) of LCD to GPIO13 of Raspberry Pi.
- Connect pin12 (D5) of LCD to GPIO6 of Raspberry Pi.
- Connect Pin13 (D6) of LCD to GPIO5 of Raspberry Pi.
- Connect Pin14 (D7) of LCD to GPIO21 of Raspberry Pi.

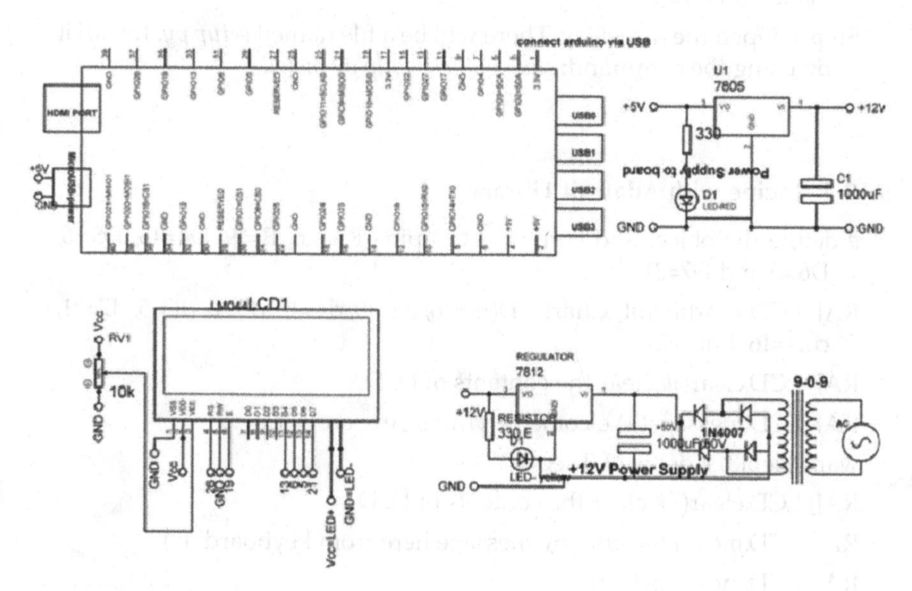

FIGURE 5.28
Circuit diagram for interfacing of LCD with Raspberry Pi.

5.19.1 Adafruit Library for LCD

Execute the commands in the Raspberry Pi terminal to install and update the libraries.

1. sudo apt-get update
2. sudo apt-get install build-essential python-dev python-smbus python-pip git
3. sudo pip install RPi. GPIO

Ignore the warnings about any dependencies. Adafruit provides a library to operate LCD in the 4-bit mode.

Steps to install library on Raspberry Pi:

Step 1: Install Git on Raspberry Pi by using the command: sudo*apt-get install git.*

Git helps to clone the project files on github and use it on Raspberry Pi.

Step 2: To make a clone of the project file on Pi, execute the command: *sudogit clone git://github.com/adafruit/Adafruit_Python_CharLCD*

Step 3: Change the directory line with the command: *cd Adafruit_Python_CharLCD*

Step 4: Open the directory. There will be a file named *setup.py*. Install it by using the command: *sudo python setup.py install*

5.19.2 Recipe with Adafruit Library

```
# define the object and connect LCD pins RS=26, E=19, D4=13, D5=6, D6=5 and D7=21
RAJ_LCD = Adafruit_CharLCD(rs=26, en=19, d4=13, d5=6, d6=5, d7=21, cols=16, lines=2)
RAJ_LCD.clear()# clear the contents of LCD
RAJ_LCD.message('Welcome') # print string on LCD
wait.sleep(3) # delay of 3 sec
RAJ_LCD.clear() # clear the contents of LCD
RAJ_LCD.message(' put any message here from keyboard\n')
RAJ_LCD.message(text)
wait.sleep(5) # delay of 5 sec
RAJ_LCD.clear() # clear the contents of LCD
```

5.19.3 Recipe without Library

```
importRPi.GPIO as RAJ
import time as wait
RAJ_LCD_RS = 26 # RS pin of LCD
RAJ_LCD_E = 19 # E pin of LCD
RAJ_LCD_D4 = 13 # D4 pin of LCD
RAJ_LCD_D5 = 6 # D5 pin of LCD
RAJ_LCD_D6 = 5 # D6 pin of LCD
RAJ_LCD_D7 = 21 # D7 pin of LCD
RAJ_LCD_WIDTH = 16 # total char are 16
RAJ_LCD_CHR = True # assign Boolean
RAJ_LCD_CMD = False # assign Boolean
RAJ_LCD_LINE_1 = 0×80 # RAM address of LCD for first line
RAJ_LCD_LINE_2 = 0×C0 # RAM address of LCD for second line
RAJ_E_PULSE = 0.0005 # give time constant
RAJ_E_DELAY = 0.0005 # give time constant

def main():
RAJ.setwarnings(False) # remove the warnings
RAJ.setmode(RAJ.BCM) # Use BCM GPIO numbers
RAJ.setup(RAJ_LCD_E, RAJ.OUT) # setup enable pin as output
RAJ.setup(RAJ_LCD_RS, RAJ.OUT) # setup RS pin as output
RAJ.setup(RAJ_LCD_D4, RAJ.OUT) # setup D4 pin as output
RAJ.setup(RAJ_LCD_D5, RAJ.OUT) # setup D5 pin as output
RAJ.setup(RAJ_LCD_D6, RAJ.OUT) # setup D6 pin as output
RAJ.setup(RAJ_LCD_D7, RAJ.OUT) # setup D7 pin as output
RAJ_lcd_init() # initialise LCD
while True:
RAJ_lcd_string("R Singh ",LCD_LINE_1) # print some text to ROW 1
  of LCD
RAJ_lcd_string(" Presents ",LCD_LINE_2) # print some text to ROW2
  of LCD
wait.sleep(3) # delay of 3 Sec
RAJ_lcd_string("123456789012345",LCD_LINE_1)
RAJ_lcd_string("ABCDEFGHIJK",LCD_LINE_2)
wait.sleep(3)# delay of 3 Sec
```

```
defRAJ_lcd_init():
RAJ_lcd_display(0x28,RAJ_LCD_CMD) # choose 4 bit mode and rows
RAJ_lcd_display(0x0C,RAJ_LCD_CMD) # To ON the display, no cursor
  blink
RAJ_lcd_display(0x01,RAJ_LCD_CMD) # Clear the contents of LCD
Wait.sleep(RAJ_E_DELAY) # delay

defRAJ_lcd_display(bits, mode):
RAJ.output(RAJ_LCD_RS, mode) # set Mode
RAJ.output(RAJ_LCD_D4, False) # make D4 pin of LCD LOW
RAJ.output(RAJ_LCD_D5, False) # make D5 pin of LCD LOW
RAJ.output(RAJ_LCD_D6, False) # make D6 pin of LCD LOW
RAJ.output(RAJ_LCD_D7, False) # make D7 pin of LCD LOW
if bits&0x10==0x10:
RAJ.output(RAJ_LCD_D4, True) # make D4 pin of LCD HIGH
if bits&0x20==0x20:
RAJ.output(RAJ_LCD_D5, True) # make D5 pin of LCD HIGH
if bits&0x40==0x40:
RAJ.output(RAJ_LCD_D6, True) # make D6 pin of LCD HIGH
if bits&0x80==0x80:
RAJ.output(RAJ_LCD_D7, True) # make D7 pin of LCD HIGH

RAJ_lcd_toggle_enable()
defRAJ_lcd_toggle_enable():
wait.sleep(RAJ_E_DELAY) # delay
RAJ.output(RAJ_LCD_E, True) # make E pin to HIGH
wait.sleep(RAJ_E_PULSE) # delay
RAJ.output(RAJ_LCD_E, False) # make E pin to LOW
wait.sleep(RAJ_E_DELAY) # delay

defRAJ_lcd_string(message,line):
message = message.ljust(RAJ_LCD_WIDTH," ")
RAJ_lcd_display(line, RAJ_LCD_CMD)
for i in range(RAJ_LCD_WIDTH):
RAJ_lcd_display(ord(message[i]),RAJ_LCD_CHR)

if __name__ == '__main__':
try:
main()
```

```
exceptKeyboardInterrupt:
pass
finally:
RAJ_lcd_display(0x01, RAJ_LCD_CMD)
RAJ.cleanup() # clean all GPIOs
```

5.20 Interfacing LCD with Raspberry Pi in I2C Mode

The circuit is comprised of Raspberry Pi, LCD, and a power supply. An I2C driver is connected with LCD. Make the component connections as shown in Figure 5.29.

Step 1: Interface the LCD with Raspberry Pi

Connections:

- Connect pin1 (Vss) and pin16 (LED-) of LCD and pin (GND) of the I2C driver to ground.
- Connect pin2 (Vcc) and pin15 (LED+) of LCD and pin (Vcc) of the I2C driver to +5 V DC.

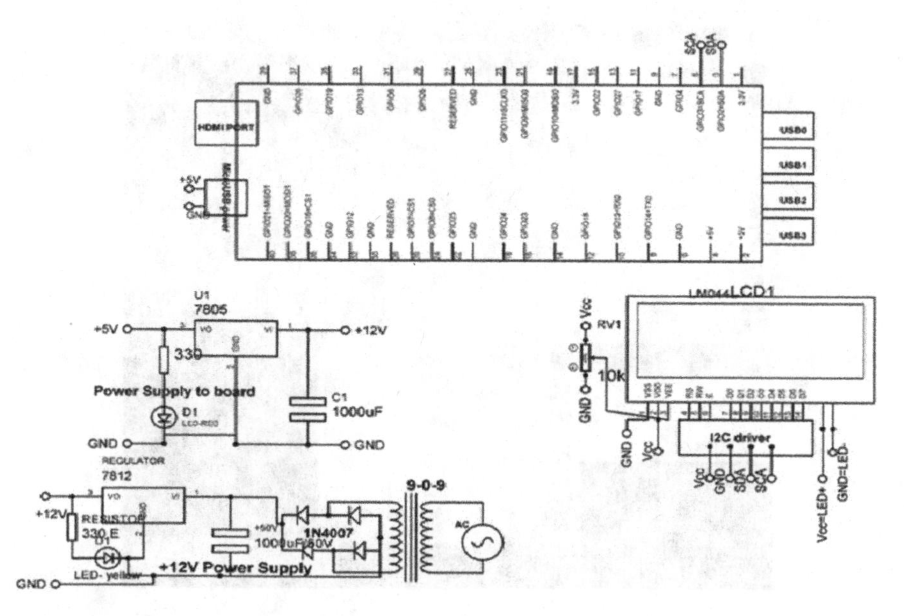

FIGURE 5.29.
Circuit diagram for LCD interfacing in I2C mode.

- Connect pin3 (V_{EE}) of LCD to the output of a 10 K potentiometer and connect the other two terminals of the potentiometer to +5 V DC and ground.
- Connect pin(SDA) of the I2C driver LCD to GPIO3 of Raspberry Pi.
- Connect pin(SCA) of the I2C driver LCD to GPIO5 of Raspberry Pi.

Step 2: Download the Python Script by using:

wget https://bitbucket.org/MattHawkinsUK/rpispy-misc/raw/master/python/lcd_i2c.py

Step 3: Enable the I2C Interface on Raspberry Pi by following the path Menu>Preferences > Raspberry Pi Configuration (Figure 5.30). Then select the "Interfaces" tab and set I2C to "Enabled" (Figure 5.31).

Click the "OK" button. If prompted to reboot, select "Yes" so that the changes can be taken effectively (Figure 5.32). The Raspberry Pi will reboot, and the I2C interface is now enabled.

Step 4: Open the LCD Script with the command: *sudonano lcd_i2c.py*

Step 5: Run the script with the command: *sudo python lcd_i2c.py*

FIGURE 5.30
Setting on terminal.

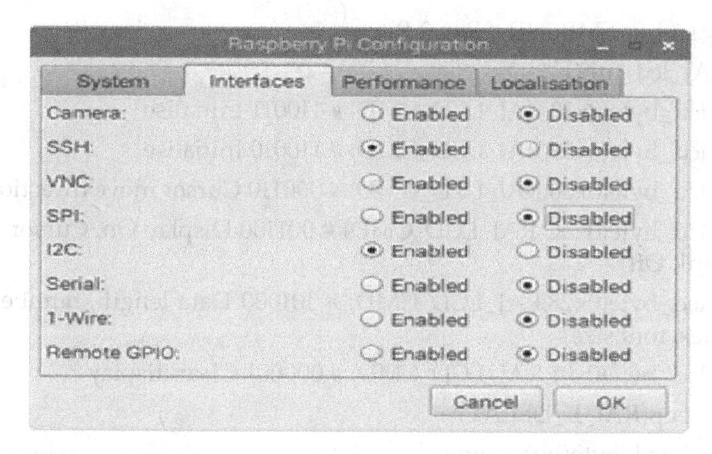

FIGURE 5.31
Enable I2C.

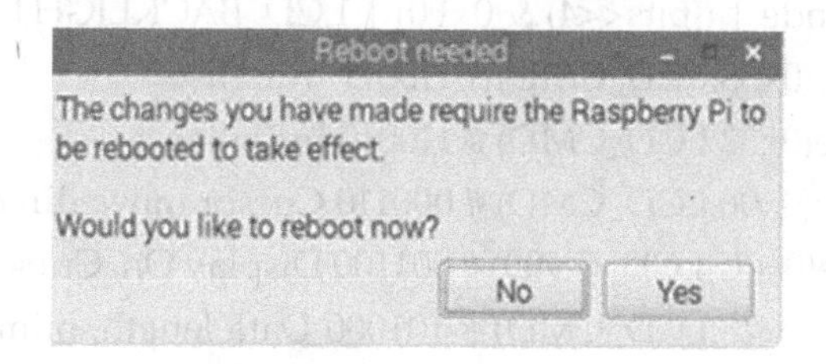

FIGURE 5.32
Reboot.

5.20.1 Recipe to Interface LCD in I2C Mode

```
importsmbus
import time as wait
RAJ_I2C_ADDR = 0x24 # address of I2C device
RAJ_LCD_WIDTH = 16 # number of char per line
RAJ_LCD_CHR = 1 # Mode - Sending data
RAJ_LCD_CMD = 0 # Mode - Sending command
RAJ_LCD_LINE_1 = 0x80 # LCD RAM address for the 1st line
RAJ_LCD_LINE_2 = 0xC0 # LCD RAM address for the 2nd line
RAJ_LCD_LINE_3 = 0x94 # LCD RAM address for the 3rd line
RAJ_LCD_LINE_4 = 0xD4 # LCD RAM address for the 4th line
RAJ_LCD_BACKLIGHT = 0x08
RAJ_ENABLE = 0b00000100 # LCD enable(E) bit
RAJ_E_PULSE = 0.0005
RAJ_E_DELAY = 0.0005
```

```
bus = smbus.SMBus(1) # Rev 2 Pi uses 1
defRAJ_lcd_init():
RAJ_lcd_byte(0x33,RAJ_LCD_CMD) # 110011 Initialise
RAJ_lcd_byte(0x32,RAJ_LCD_CMD) # 110010 Initialise
RAJ_lcd_byte(0x06,RAJ_LCD_CMD) # 000110 Cursor move direction
RAJ_lcd_byte(0x0C,RAJ_LCD_CMD) # 001100 Display On, Cursor Off,
   Blink Off
RAJ_lcd_byte(0x28,RAJ_LCD_CMD) # 101000 Data length, number of
   lines, font size
RAJ_lcd_byte(0x01,RAJ_LCD_CMD) # 000001 Clear display
wait.sleep(RAJ_E_DELAY)
defRAJ_lcd_byte(bits, mode):
bits_high = mode | (bits & 0xF0) | LCD_BACKLIGHT
bits_low = mode | ((bits<<4) & 0xF0) | LCD_BACKLIGHT
RAJ_lcd_byte(0x33,LCD_CMD) # 110011 Initialise
RAJ_lcd_byte(0x32,LCD_CMD) # 110010 Initialise
RAJ_lcd_byte(0x06,LCD_CMD) # 000110 Cursor move direction
RAJ_lcd_byte(0x0C,LCD_CMD) # 001100 Display On, Cursor Off, Blink Off
RAJ_lcd_byte(0x28,LCD_CMD) # 101000 Data length, number of lines,
   font size
RAJ_lcd_byte(0x01,LCD_CMD) # 000001 Clear display
wait.sleep(RAJ_E_DELAY)
bus.write_byte(I2C_ADDR, bits_high)
RAJ_lcd_toggle_enable(bits_high)
bus.write_byte(RAJ_I2C_ADDR, bits_low)
RAJ_lcd_toggle_enable(bits_low)
defRAJ_lcd_toggle_enable(bits):
wait.sleep(RAJ_E_DELAY) # wait
bus.write_byte(I2C_ADDR, (bits | RAJ_ENABLE))
wait.sleep(RAJ_E_PULSE) # wait
bus.write_byte(I2C_ADDR,(bits & ~RAJ_ENABLE))
wait.sleep(RAJ_E_DELAY) # wait
deflcd_string(message,line):
message = message.ljust(RAJ_LCD_WIDTH," ")
RAJ_lcd_byte(line, RAJ_LCD_CMD)
for i in range(RAJ_LCD_WIDTH):
```

```
RAJ_lcd_byte(ord(message[i]),RAJ_LCD_CHR)
def main():
RAJ_lcd_init()
while True:
RAJ_lcd_string("I2C LCD<",RAJ_LCD_LINE_1) # print string on line1
  on LCD
RAJ_lcd_string("Connected<",RAJ_LCD_LINE_2) # print string on
  line2 on LCD
wait.sleep(3) # delay of 3 Sec
RAJ_lcd_string(">RPiSpy", RAJ_LCD_LINE_1) # print string on line1
  on LCD
RAJ_lcd_string(">    I2C LCD", RAJ_LCD_LINE_2) # print string on
  line2 on LCD
wait.sleep(3) # delay of 3 Sec
if __name__ == '__main__':
try:
main()
exceptKeyboardInterrupt:
pass
finally:
RAJ_lcd_byte(0x01, RAJ_LCD_CMD)
```

5.21 Interfacing of DHT11 Sensor with Raspberry Pi

The DHT11 sensor works with a one-wire system. The temperature and humidity values are measured by the sensor, and then this data is communicated serially through the output pin. It is a four-pin device. A resistor of value 4.4–10 K needs to be placed between pin 1 (3.3 V) and pin 2 (Data). The three-pin device has a built-in resistor (Figure 5.33).

The Adafruit DHT11 library can be used for DHT11, DHT22, or any other one-wire temperature sensor. The procedure to install the DHT11 library is similar to the steps followed for installing the LCD library.

Steps to interface DHT11

1. Update the package lists and install the Python libraries with the help of commands:

   sudo apt-get update

   sudo apt-get install build-essential python-dev

FIGURE 5.33
View of DHT11 sensor.

2. Download the Adafruit library by using the command:
   git clone https://github.com/adafruit/Adafruit_Python_DHT.git
   cdAdafruit_Python_DHT
3. Install the library for Python 2 and Python 3 by using the command:
   sudo python setup.py install
   sudo python3 setup.py install
4. Connect the components as follows:

Connections:

- Connect pin1 (Vss), pin5 (RW) and pin16 (LED-) of LCD to ground.
- Connect pin2 (Vcc), and pin15 (LED+) of LCD to +5 V DC.
- Connect pin3 (V_{EE}) of LCD to output of 10 K potentiometer and connect other two terminals of potentiometer to +5 V DC and ground.
- Connect pin4 (RS) of LCD to GPIO26 of Raspberry Pi.
- Connect pin6 (E) of LCD to GPIO19 of Raspberry Pi.
- Connect pin11 (D4) of LCD to GPIO13 of Raspberry Pi.
- Connect pin12 (D5) of LCD to GPIO6 of Raspberry Pi.
- Connect pin13 (D6) of LCD to GPIO5 of Raspberry Pi.
- Connect pin14 (D7) of LCD to GPIO21 of Raspberry Pi.
- Connect pin (Vcc) and pin (GND) of DHT11 to +5 V and ground, respectively.
- Connect output pin of DHT11 to GPIO17 of Raspberry Pi.

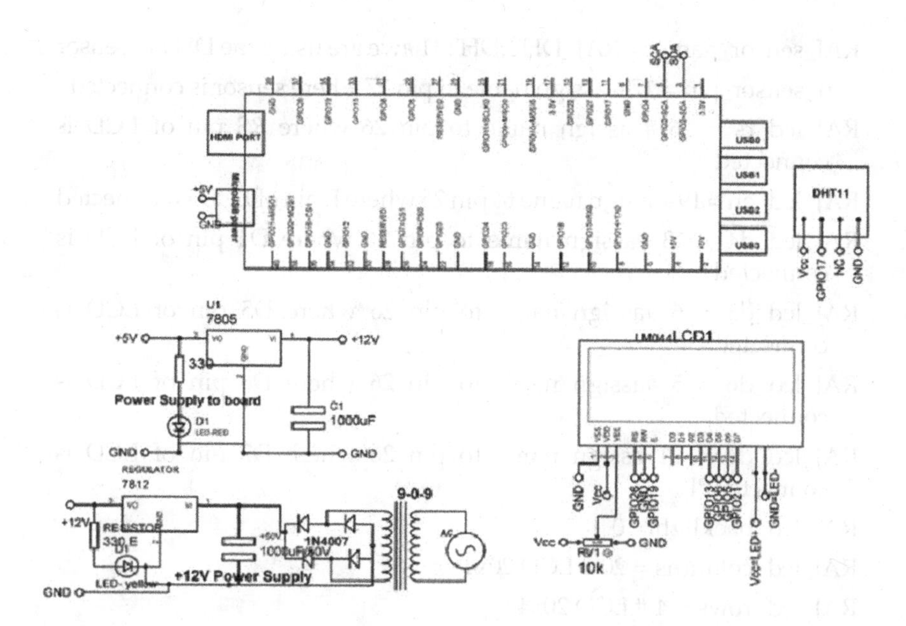

FIGURE 5.34
Circuit diagram for the interfacing of DHT11S.

Figure 5.34 shows the circuit diagram for the interfacing of LCD and DHT11 with Raspberry Pi.

5.21.1 Recipe to Read DHT11 Sensor

```
    importAdafruit_DHT as RAJ_DHT
    import time as wait
    RAJ_sensor = RAJ_DHT.DHT11
    RAJ_gpio = 17 # assign variable to pin 17
    HUM, TEMP = RAJ_DHT.read_retry(RAJ_sensor, RAJ_gpio)
    if humidity is not None and temperature is not None:
print('TEMP={0:0.1f}*C HUM={1:0.1f}%'.format(TEMP, HUM))
wait.sleep(0.2)
    else:
    print('try again for reading from DHT11')
    wait.sleep(0.2)
```

5.21.2 Recipe to Read DHT11 Sensor and Display Data on LCD

```
import time as wait #import time library
importAdafruit_CharLCD as RAJ_LCD #Import the LCD library
importAdafruit_DHTas RAJ_DHT #Import DHT library
```

```
RAJ_sensor_name  = RAJ_DHT.DHT11 #we are using the DHT11 sensor
RAJ_sensor_pin = 17 # assign name to pin 17 where sensor is connected
RAJ_lcd_rs  = 26 # assign name to pin 26 where RS pin of LCD is
   connected
RAJ_lcd_en = 19 #assign name to pin 26 where E pin of LCD is connected
RAJ_lcd_d4  = 13 #assign name to pin 26 where D4 pin of LCD is
   connected
RAJ_lcd_d5  = 6 #assign name to pin 26 where D5 pin of LCD is
   connected
RAJ_lcd_d6  = 5 #assign name to pin 26 where D6 pin of LCD is
   connected
RAJ_lcd_d4  = 21 assign name to pin 26 where D7 pin of LCD is
   connectedPI
RAJ_lcd_backlight= 0
RAJ_lcd_columns = 20 # LCD 20x4
RAJ_lcd_rows = 4 # LCD 20x4
RAJ_lcd = RAJ_LCD_CharLCD(RAJ_lcd_rs, RAJ_lcd_en, RAJ_lcd_d4,
   RAJ_lcd_d5, RAJ_lcd_d6, RAJ_lcd_d4, RAJ_lcd_columns, RAJ_lcd_
   rows, RAJ_lcd_backlight)
#provide pin details to LCD library
RAJ_lcd.message("VALUE FROM DHT11") #print message on LCD
wait.sleep(2) # delay of 2 sec
while 1:
HUM, TEMP = Adafruit_DHT.read_retry(RAJ_sensor_name, RAJ_
   sensor_pin)
   RAJ_lcd.clear() #clear the contents of LCD
   RAJ_lcd.message ('Temp= %.1f C' % TEMP) # print temp value on LCD
   RAJ_lcd.message ('\nHum = %.1f %%' % HUM) print hum value on LCD
print('TEMP={0:0.1f}*C HUM={1:0.1f}%'.format(TEMP, HUM))
   wait.sleep(2) #delay of 2 sec
```

5.22 Interfacing of Ultrasonic Sensor with Raspberry Pi

An ultrasonic sensor is used to measure the distance of an object from the sensor. It consists of an ultrasonic transmitter, a control circuit, and an ultrasonic receiver. It has four pins, namely, VCC, TRIG (Trigger), ECHO (Echo),

FIGURE 5.35
Ultrasonic sensor.

and GND. The ultrasonic transmitter in the sensor generates a 40 KHz ultrasound, which propagates in the air and reflects back if any obstacle occurs in its path. The distance is calculated on the basis of a time interval between the transmitted signal and receiving it back to echo the pin of the sensor. To generate the ultrasonic signal trigger, the pin needs to hold "HIGH" for a minimum duration of 10 μS. Figure 5.35 shows the ultrasonic sensor.

For example, if the time for which ECHO is HIGH is 588 μS, then the distance can be calculated with the help of the speed of sound, which is equal to 340 m/s.

$$\text{Distance} = \text{Velocity of Sound}/(\text{Time}/2)$$

$$= 340\ \text{m/s}\ /\left(588\ \mu\text{S}/2\right) = 10\ \text{cm}.$$

This formula needs to be used in the program to calculate the distance.

To understand the working of the ultrasonic sensor, a simple circuit can be designed. It is comprised of a Raspberry Pi, a power supply, an ultrasonic sensor, and an LCD. Connect the components as follows:

Connections:

- Connect pin1 (Vss), pin5 (RW), and pin16 (LED–) of LCD to ground.
- Connect pin2 (Vcc) and pin15 (LED+) of LCD to +5 V DC.
- Connect pin3 (V_{EE}) of LCD to output of 10 K potentiometer, and connect the other two terminals of the potentiometer to +5 V DC and ground.
- Connect pin4 (RS) of LCD to GPIO26 of Raspberry Pi.
- Connect pin6 (E) of LCD to GPIO19 of Raspberry Pi.
- Connect pin11 (D4) of LCD to GPIO13 of Raspberry Pi.

- Connect pin12 (D5) of LCD to GPIO6 of Raspberry Pi.
- Connect pin13 (D6) of LCD to GPIO5 of Raspberry Pi.
- Connect pin14 (D7) of LCD to GPIO21 of Raspberry Pi.
- Connect pin (Vcc) and pin (GND) of ultrasonic to +5 V and ground, respectively.
- Connect pin (TRIG) of ultrasonic sensor to GPIO20 of Raspberry Pi.
- Connect pin (ECHO) of ultrasonic sensor to GPIO21 of Raspberry Pi.

Figure 5.36 shows the circuit diagram for the interfacing of the LCD and ultrasonic sensor with Raspberry Pi.

5.22.1 Recipe to Read Ultrasonic Sensor and Display Data on LCD

```
importRPi.GPIO as ANITA
import time as wait
importAdafruit_CharLCD as LCD #Import LCD library
RAJ_lcd_rs = 26 #assign name to pin 26 where RS pin of LCD is
  connected
RAJ_lcd_en = 19 #assign name to pin 26 where E pin of LCD is connected
RAJ_lcd_d4 = 13 #assign name to pin 26 where D4 pin of LCD is
  connected
```

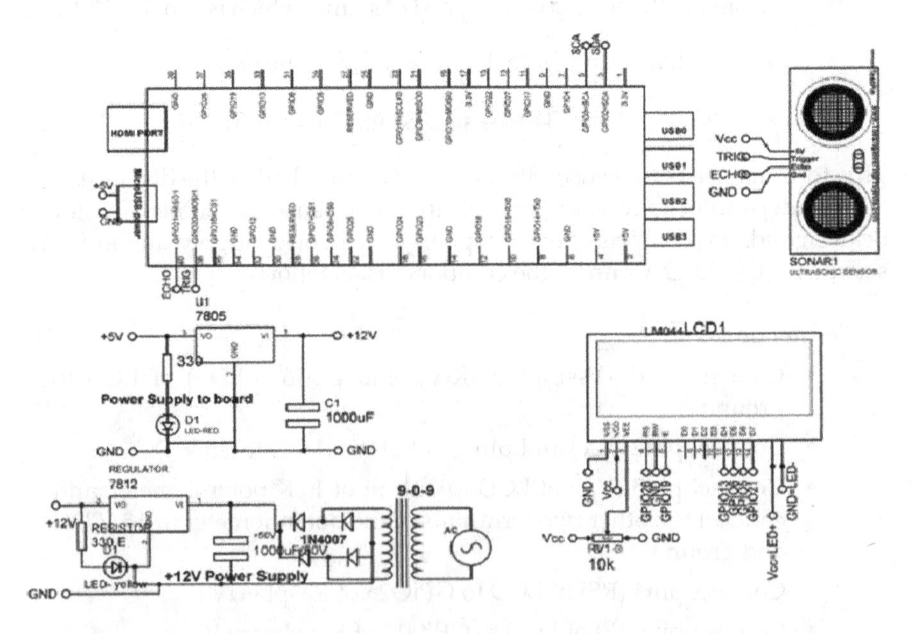

FIGURE 5.36
Circuit diagram for interfacing of ultrasonic sensor with Raspberry Pi.

```
RAJ_lcd_d5 = 6 #assign name to pin 26 where D5 pin of LCD is
  connected
RAJ_lcd_d6 = 5 #assign name to pin 26 where D6 pin of LCD is
  connected
RAJ_lcd_d7 = 21 #assign name to pin 26 where D7 pin of LCD is
  connected
RAJ_lcd_backlight = 0 #assign 0 input to backlight
RAJ_lcd_columns = 20 #choose 20x4 LCD
RAJ_lcd_rows = 4 #fochoose 20x4 LCD
RAJ_lcd = LCD.Adafruit_CharLCD(RAJ_lcd_rs, RAJ_lcd_en, RAJ_
  lcd_d4, RAJ_lcd_d5, RAJ_lcd_d6, RAJ_lcd_d4, RAJ_
  lcd_columns, RAJ_lcd_rows, RAJ_lcd_backlight) #supply details
  to LCD library
GPIO.setmode(GPIO.BCM) #set mode of GPIO to BCM
RAJ_GPIO_TRIGGER = 20 # assign 20 pin to trigger pin of sensor
RAJ_GPIO_ECHO = 21 # assign 21 pin to ECHO pin of sensor
ANITA.setup(RAJ_GPIO_TRIGGER, ANITA.OUT) # set direction of
  pin as #OUTPUT
ANITA.setup(RAJ_GPIO_ECHO, ANITA.IN) # set direction of pin as
  OUTPUT
def distance():
ANITA.output(RAJ_GPIO_TRIGGER, True) # make 20 pin to HIGH
wait.sleep(0.00001) # delay of 10 uSec
ANITA.output(RAJ_GPIO_TRIGGER, False) # make 20 pin to LOW
Start_Time = wait.time()
Stop_Time = wait.time()
whileANITA.input(RAJ_GPIO_ECHO) == 0:
        Start_Time = wait.time()
whileANITA.input(RAJ_GPIO_ECHO) == 1:
        Stop_Time = wait.time()
Time_Elapsed = Stop_Time–Start_Time # record elapsed time
RAJ_distance = (Time_Elapsed * 34300) / 2
returnRAJ_distance
if __name__ == '__main__':
try:
lcd.message('DISTANCE:') #print string on LCD
wait.sleep(2) #wait for 2 secs
```

```
while True:
        RAJ_dist = distance()
        print ("DISTANCE = %.1f cm" % RAJ_dist)
        RAJ_lcd.clear() #clear the contents of LCD
        RAJ_lcd.message ('DIST = %.1f cm' % RAJ_dist)
        wait.sleep(1)
exceptKeyboardInterrupt:
print("stop measurement")
ANITA.cleanup()
```

5.23 Interfacing of Camera with Raspberry Pi

A camera is used to take still pictures or videos. The interfacing of Raspberry Pi with a camera module can be done with Python and Picamera. Switch off the Raspberry Pi then connect the camera to the camera port and switch on the Raspberry Pi.

5.23.1 Configuring the Camera with Raspberry Pi

1. Connect the camera and start up the Raspberry Pi.
2. Open the configuration tool from the main menu (Figure 5.37).
3. Enable the camera software and reboot the Raspberry Pi (Figure 5.38).

5.23.2 Capturing the Image with Pi Camera

1. Make a new editable file using the command *sudonano image_RAJ.py* on Pi terminal

```
frompicamera import PiCamera
import time as wait
RAJ_camera = PiCamera()
RAJ_camera.start_preview()
wait.sleep(10)
RAJ_camera.stop_preview()
```

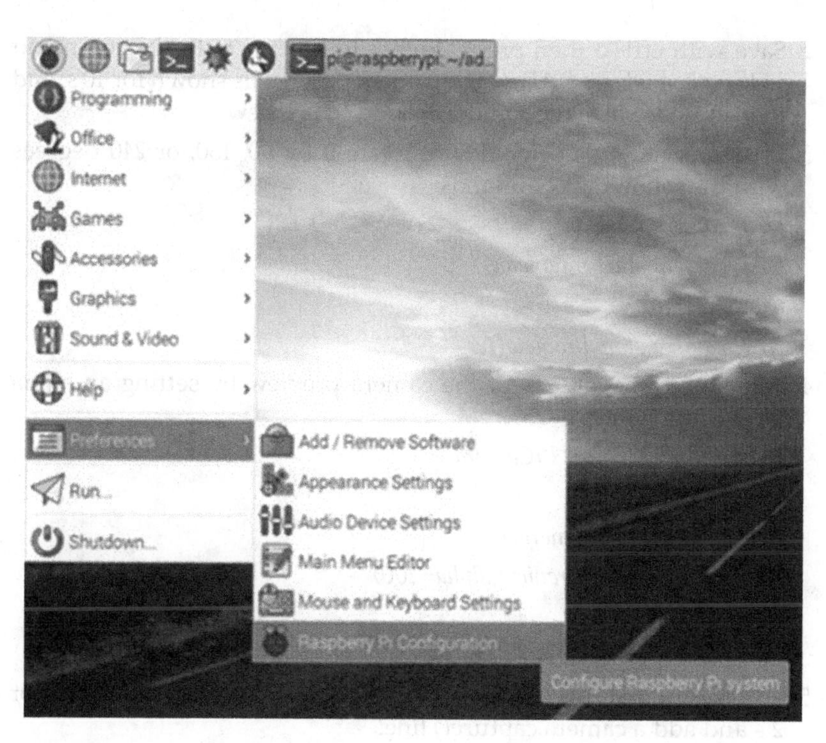

FIGURE 5.37
Configuring the Raspberry Pi.

Raspberry Pi Configuration

System | Interfaces | Performance | Localisation

Camera:	⦿ Enabled	○ Disabled
SSH:	⦿ Enabled	○ Disabled
VNC:	○ Enabled	⦿ Disabled
SPI:	○ Enabled	⦿ Disabled
I2C:	○ Enabled	⦿ Disabled
Serial:	○ Enabled	⦿ Disabled
1-Wire:	○ Enabled	⦿ Disabled
Remote GPIO:	○ Enabled	⦿ Disabled

Cancel | OK

FIGURE 5.38
Enable camera.

2. Save with ctrl+o then press "Enter" to execute with the command *sudo python image_RAJ.py*. The camera preview is shown for 10 s, and then closed. Move the camera around to preview.
3. If the preview is upside-down, rotate it by 90, 180, or 240 degrees with the following commands:

```
RAJ_camera.rotation = 180
RAJ_camera.start_preview()
wait.sleep(10)
RAJ_camera.stop_preview()
```

4. Alter the transparency of the camera preview by setting an alpha level (from values 0–255):

```
frompicamera import PiCamera
import time as wait
RAJ_camera = PiCamera()
RAJ_camera.start_preview(alpha=200)
sleep(10)
camera.stop_preview()
```

5. To take still pictures, amend the code to reduce the sleep at least for 2 s and add a camera.capture() line:

```
camera.start_preview()
sleep(5)
RAJ_camera.capture('/home/pi/Desktop/image.jpg')
RAJ_camera.stop_preview()
```

6. Run the code and see the camera preview open for 5 s before capturing a still picture.
7. The picture is on the desktop. Double-click the file icon to open it.
8. Now try adding a loop to take five pictures:

```
RAJ_camera.start_preview()
for i in range(5):
wait.sleep(5)
RAJ_camera.capture('/home/pi/Desktop/image%s.jpg' % i)
RAJ_camera.stop_preview()
```

   The variable "I" contains the current iteration number, from 0 to 4, so the images will be saved as image0.jpg, image1.jpg, and so on.
9. Pictures can be resized by using libraries like PIL and OpenCV. This can be done with the resize parameter of the capture() methods:

```
import time as wait
```

importpicamera
withpicamera. PiCamera() as RAJ_camera:
RAJ_camera.resolution = (1024, 468)
RAJ_camera.start_preview()
wait.sleep(2)
RAJ_camera.capture('image.jpg', resize=(320, 240))

5.23.3 Capturing the Video with Pi Camera

1. Change the code for the image to replace capture() with start_recording() and stop_recording():
   RAJ_camera.start_preview()
   RAJ_camera.start_recording('/home/pi/video.h264')
   wait.sleep(10)
   RAJ_camera.stop_recording()
   RAJ_camera.stop_preview()
2. Run the code; it will record video for 10 s and then close the preview. To play the video, open a terminal window by clicking the black monitor icon in the taskbar (Figure 5.39).
3. Open the terminal and type the command *omxplayer video.h264* and press "Enter" to play the video.
4. The resolution of the camera can be configured with commands:
   RAJ_camera.resolution = (2592, 1944)
   RAJ_camera.framerate = 15
   RAJ_camera.start_preview()
   wait.sleep(5)
   RAJ_camera.capture('/home/pi/Desktop/max.jpg')
   RAJ_camera.stop_preview()
   "The minimum resolution allowed is 64 × 64."
5. Add text to the image with annotate_text by the commands:
   RAJ_camera.start_preview()
   RAJ_camera.annotate_text = "Hello world!"

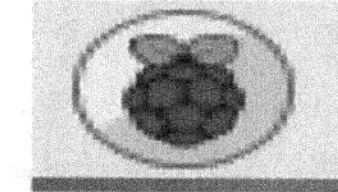

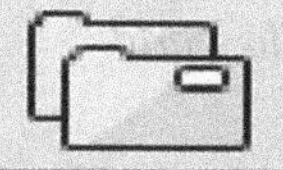

FIGURE 5.39
Taskbar.

wait.sleep(5)
RAJ_camera.capture('/home/pi/Desktop/text.jpg')
RAJ_camera.stop_preview()

6. The default brightness is "50"; it can be altered by the commands:
   RAJ_camera.start_preview()
   RAJ_camera.brightness = 40
   wait.sleep(5)
   RAJ_camera.capture('/home/pi/Desktop/bright.jpg')
   RAJ_camera.stop_preview()
7. The brightness can be altered by applying the loop:
   RAJ_camera.start_preview()
   for i in range(100):
   RAJ_camera.annotate_text = "Brightness: %s" % i
   RAJ_camera.brightness = i
   wait.sleep(0.1)
   RAJ_camera.stop_preview()
8. The contract can also be altered by using commands:
   RAJ_camera.start_preview()
   for i in range(100):
   RAJ_camera.annotate_text = "Contrast: %s" % i
   RAJ_camera.contrast = i
   wait.sleep(0.1)
   RAJ_camera.stop_preview()
9. The annotation text size (value from 6 to 160) with default value 32 can be added with the command:
   RAJ_camera.annotate_text_size = 50
10. The annotation colors can be altered by using the commands:
    frompicamera import PiCamera, Color
11. Then other colors can also be amended by using the commands:
    RAJ_camera.start_preview()
    RAJ_camera.annotate_background = Color('blue')
    RAJ_camera.annotate_foreground = Color('yellow')
    RAJ_camera.annotate_text = "Hello world "
    wait.sleep(5)
    RAJ_camera.stop_preview()

5.23.4 Motion Detector and Capturing the Image

To understand the working of the camera, a simple circuit of capturing the image on detecting a motion is discussed here. An LED will glow on detection of a motion, or a buzzer can be put in place of an LED. The system is comprised of a Raspberry Pi, a PIR motion sensor, Pi camera, an LED, and a power supply.

Connections:

- Connect camera to Raspberry Pi as described in Section 5.23.
- Connect pin (Vcc) and pin (GND) of PIR sensor to +5 V and ground, respectively.
- Connect pin (OUT) of PIR sensor to GPIO21 of Raspberry Pi.
- Connect anode of LED to GPIO3 of Raspberry Pi through a 10 K resistor and cathode of LED to ground.

Figure 5.40 shows the circuit diagram for interfacing of the PIR sensor and camera with Raspberry Pi.

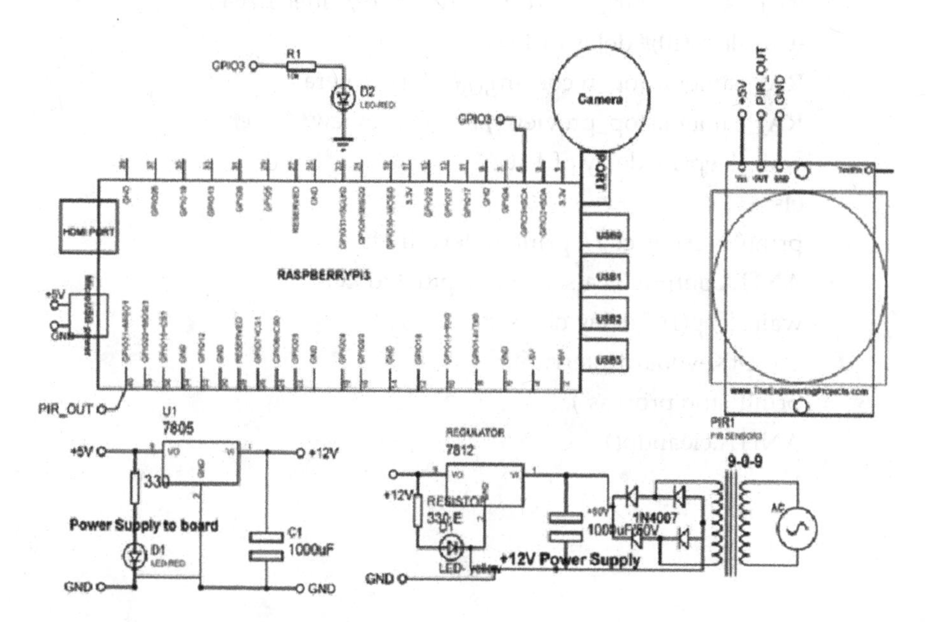

FIGURE 5.40
Circuit diagram for interfacing of PIR sensor and camera with Raspberry Pi.

5.23.4.1 Recipe to Capture Image on Motion Detection

```
importRPi.GPIO as ANITA
import time as wait
ANITA.setwarnings(False)
ANITA.setmode(ANITA.BCM)
ANITA.setup(3,ANITA.OUT) # assign pin3 as output
ANITA.setup(21,ANITA.IN) # connect Motion sensor on GPIO21
importpicamera
withpicamera.PiCamera() as RAJ_camera:
RAJ_camera.resolution = (1024, 468)
try:
while 1:
VAL=GPIO.input(21)
if(VAL==1):
print("object detected") # print on terminal
ANITA.output(3,True) # make pin 3 to HIGH
RAJ_camera.start_preview() # start camera preview
RAJ_camera.start_recording('/home/pi/video.h264')
wait.sleep(10)# delay of 1 sec
RAJ_camera.stop_recording() # stop camera
RAJ_camera.stop_preview() # stop preview
wait.sleep(1)# delay of 1 sec
else:
print("NO object") # print on terminal
ANITA.output(3,False)# make pin 3 to LOW
wait.sleep(1) # delay of 1 sec
exceptKeyboardInterrupt:
print("stop process")
ANITA.cleanup()
```

Section III

Interfacing with Raspberry Pi and Arduino

6

Raspberry Pi and Arduino

6.1 Install Arduino IDE on Raspberry Pi

The limitation with Raspberry Pi is the absence of analog ports onboard, which makes it inappropriate for the systems where analog sensors need to be read. To overcome this limitation, the Arduino integrated development environment (IDE) can be installed on Raspberry Pi, as Arduino has analog ports so these ports can be used to interface analog sensors. Installing Arduino IDE on Raspberry Pi is an easy process with simple steps. The Arduino IDE is available for most of the operating systems, but here we will see how to install it on a Raspberry Pi3 model B with running Raspbian Jessie in the graphical user interface (GUI).

1. The first requirement is an active internet connection.
2. A screen, keyboard, and mouse need to be connected with Raspberry Pi.
3. Install the latest version of Arduino IDE using apt:

   sudo apt-get update &&sudo apt-get upgrade
   sudo apt-get install arduino
4. Connect an Arduino board to the Raspberry Pi using the appropriate cable and pull down the Raspbian main Menu and select Arduino IDE under the "Electronics" head. A blank window will open. Figure 6.1 shows the blank window for Arduino IDE.
5. Click on Tools > Board > and select the appropriate board of Arduino.
6. To select the port of the Arduino that is connected, check the serial port under the "Tools" menu. The port name of Arduino is: */dev/ttyUSB0 or /dev/ttyACM0.*

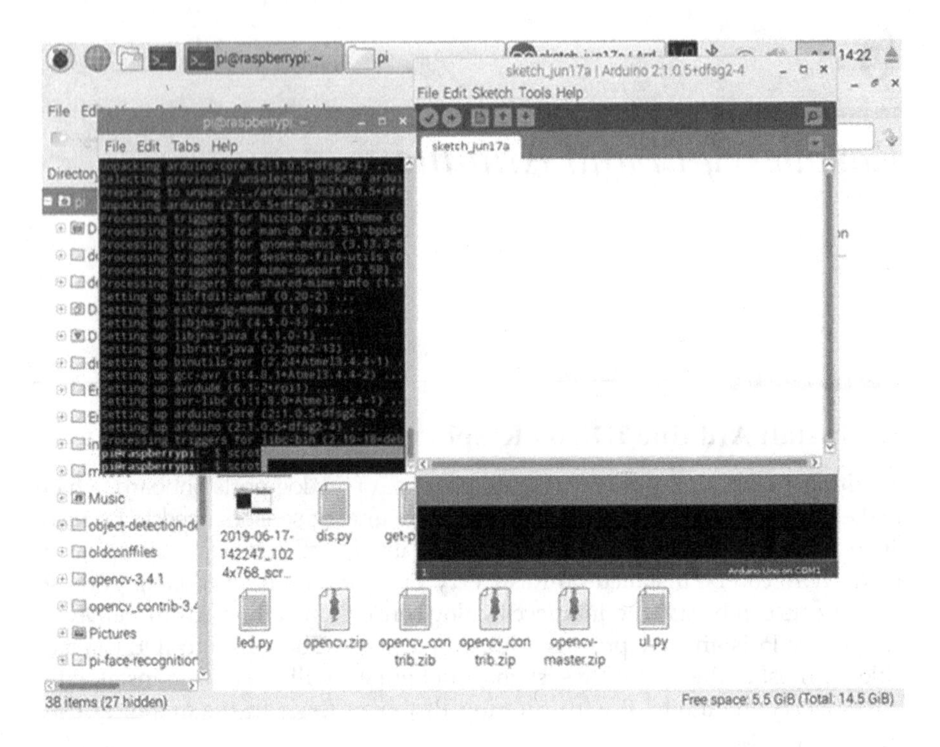

FIGURE 6.1
Blank window of Arduino IDE.

6.2 Play with Digital Sensor

After installing the Arduino IDE on Raspberry Pi, the sensors connected with the Raspberry Pi and Arduino can be read. The Arduino can simply act like the Arduino board, and Raspberry Pi acts as the computer when sensors are connected with Arduino. This can be understood with the help of a few examples.

6.2.1 PIR Sensor

The pyroelectric infrared (PIR) sensor module is used to detect motion. It is compact and easy to use. It has a Fresnel lens and motion detection circuit, which has a wide range of voltages supplied with less current drain. It has a high sensitivity and low noise. The output of the sensor is a transistor–transistor logic (TTL) active low signal. It detects motion by measuring the changes in infrared levels emitted by objects in its surroundings. This module has a detection range of 6 m and can be used in burglar alarms and the control systems. Figure 6.2 shows the block diagram of a system designed to understand the workings of a PIR sensor. It is comprised of an Arduino Uno,

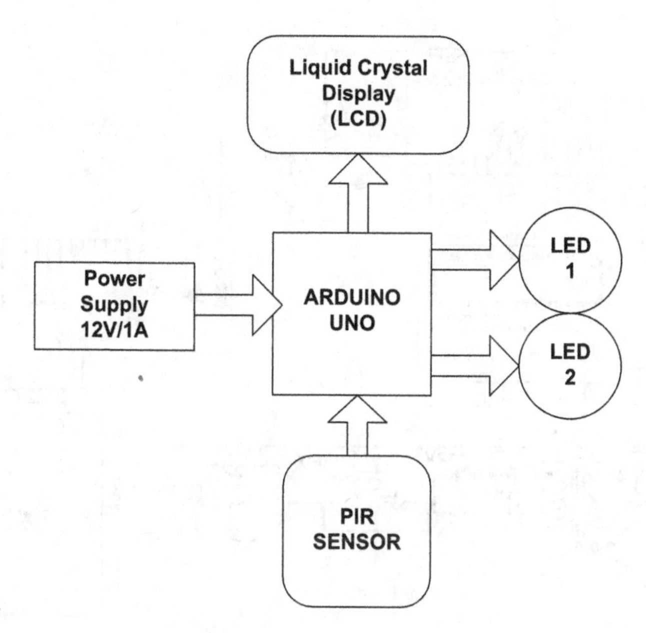

FIGURE 6.2
Block diagram for PIR interfacing with Arduino.

a PIR sensor, a liquid crystal display, and an LED. The system is designed such that "RED LED" will be "ON" if motion is detected; otherwise, "BLUE LED" will be "ON."

6.2.2 Circuit Diagram

Connect the components as shown in Figure 6.3 to check the workings of the PIR sensor. Upload the program described in Section 6.2.2 and check the workings.

PIR sensor connection

- Connect Arduino GND to PIR Module GND.
- Connect Arduino +5 V to PIR Module +.
- Connect Arduino digital pin 2 to PIR Module digital out pin.

LCD connection

- Connect Arduino digital pin (13) to RS pin(4) of LCD.
- Connect Arduino digital pin (GND) to RW pin(5) of LCD.
- Connect Arduino digital pin (12) to E pin(6) of LCD.
- Connect Arduino digital pin (11) to D4 pin(11) of LCD.
- Connect Arduino digital pin (10) to D5 pin(12) of LCD.

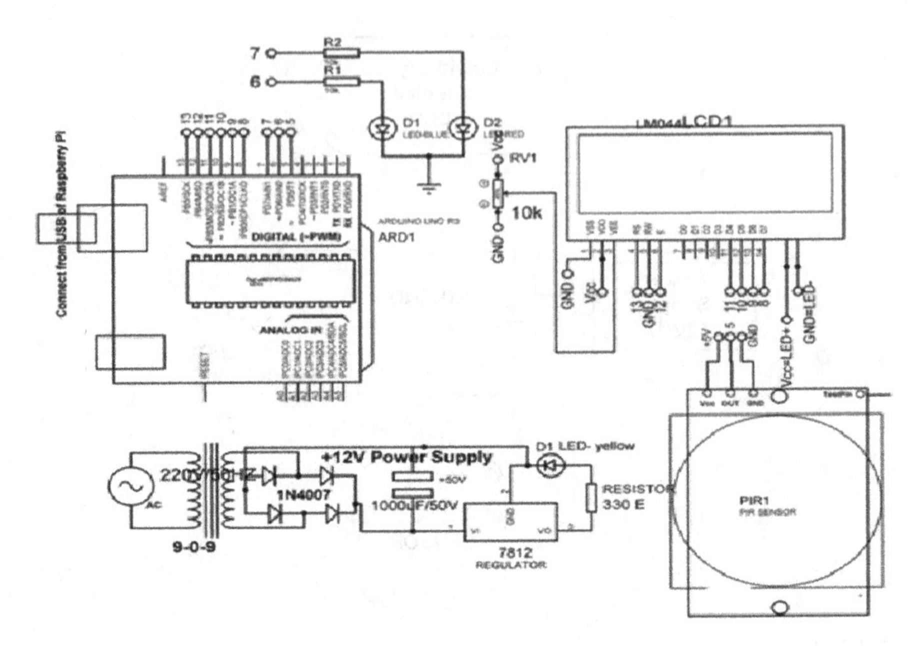

FIGURE 6.3
Circuit diagram for PIR interfacing with Arduino.

- Connect Arduino digital pin (9) to D6 pin(13) of LCD.
- Connect Arduino digital pin (8) to D7 pin(14) of LCD.

LED connection

- Connect Arduino digital pin 7 to anode of RED-LED through 330-ohm resistor.
- Connect Arduino digital pin 6 to anode BLUE-LED through 330-ohm resistor.
- Connect cathode of both LEDs to Ground.

6.2.3 Sketch

```
#include <LiquidCrystal.h> // include library of LCD
LiquidCrystallcd(13, 12, 11, 10,9, 8); // attach LCD pin RS,E,D4,D5,D6,D7
   to the given pins
int PIR_SENSOR_LOW=5; // assign pin 5 as PIR_SENSOR_LOW
int RED_LED=7; // assign pin 7 as RED_LED
int BLUE_LED=6; // // assign pin 6 as BLUE_LED
void setup()
{
```

```
pinMode(PIR_SENSOR_LOW, INPUT_PULLUP); // configure pin5 as
   an input and enable the internal pull-up resistor
pinMode(RED_LED,OUTPUT); // configure pin7 as output
pinMode(BLUE_LED,OUTPUT); // configure pin6 as output
lcd.begin(20, 4); // set up the LCD's number of columns and rows
lcd.setCursor(0, 0); // set cursor to column0 and row1
lcd.print("MOTION SENSOR BASED "); // Print a message to the LCD.
lcd.setCursor(0, 1); // set cursor to column0 and row1
lcd.print("MOTION DETECTION "); // Print a message to the LCD.
lcd.setCursor(0, 2); // set cursor to column0 and row2
lcd.print("SYSTEM AT LPU"); // Print a message to the LCD.
delay(1000);
}
void loop()
{
int PIR_SENSOR_LOW_READ = digitalRead(PIR_SENSOR_LOW);
   // read the PIR value into a variable
if (PIR_SENSOR_LOW_READ == LOW) // Read PIN 5 as LOW PIN
  {
lcd.clear(); // clear the contents of the LCD
lcd.setCursor(0, 3); // set cursor to column0 and row2
lcd.print("MOTION DETECTED "); // Print a message to the LCD.
digitalWrite(RED_LED, HIGH); // Make pin7 to HIGH
digitalWrite(BLUE_LED, LOW); // Make pin6 to LOW
delay(20); // delay of 20 mS
  }
else //otherwise
  {
lcd.clear(); // clear the contents of the LCD
lcd.setCursor(0, 3); // set cursor to column0 and row3
lcd.print("MOTION NOT DETECTED "); // Print a message to the LCD.
digitalWrite(BLUE_LED, HIGH); // Make pin 7 to HIGH
digitalWrite(RED_LED,LOW); // Low pin6 to LOW
delay(20); // delay of 20 mS
  }
}
```

6.3 Play with Analog Sensor

To read the analog sensor with Arduino, simply connect the sensor to any of the analog pins of the board. To understand the workings of the analog sensor, an example of a light-dependent resistor (LDR) is explained here. An LDR has cadmium sulphide (CdS) photoconductive cells with spectral responses. The resistance of the cells decreases with the increase in light intensity. An LDR can be used in many applications, such as smoke detection, automatic light control, batch counting, and burglar alarm systems. Figure 6.4 shows the block diagram to interface the LDR with Arduino. It is comprised of an Arduino Uno, a power supply, a liquid crystal display, and an LDR. The system is designed to display the light intensity of an LCD.

6.3.1 Circuit Diagram

Connect the components as shown in Figure 6.5 to check the workings of the LDR sensor as a simple analog sensor. This LDR has three terminals: Ground, Vcc, OUT. Upload the program described in Section 6.3.2 and check the workings.

Light sensor connection

- Connect Arduino GND to LDR module GND.
- Connect Arduino +5 V to LDR Module +.
- Connect Arduino A0 pin to OUT pin of sensor.

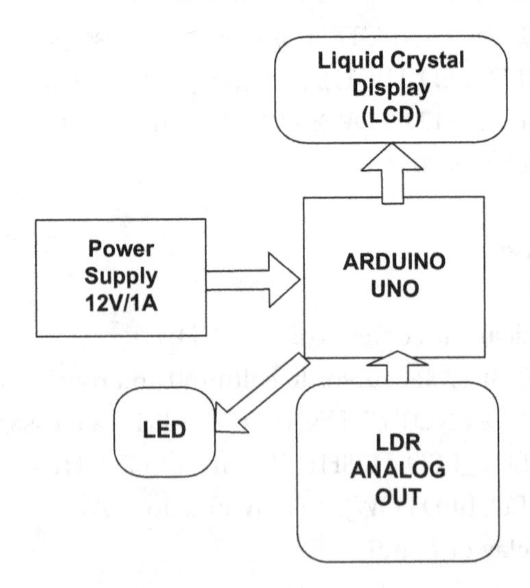

FIGURE 6.4
Block diagram to interface LDR with Arduino.

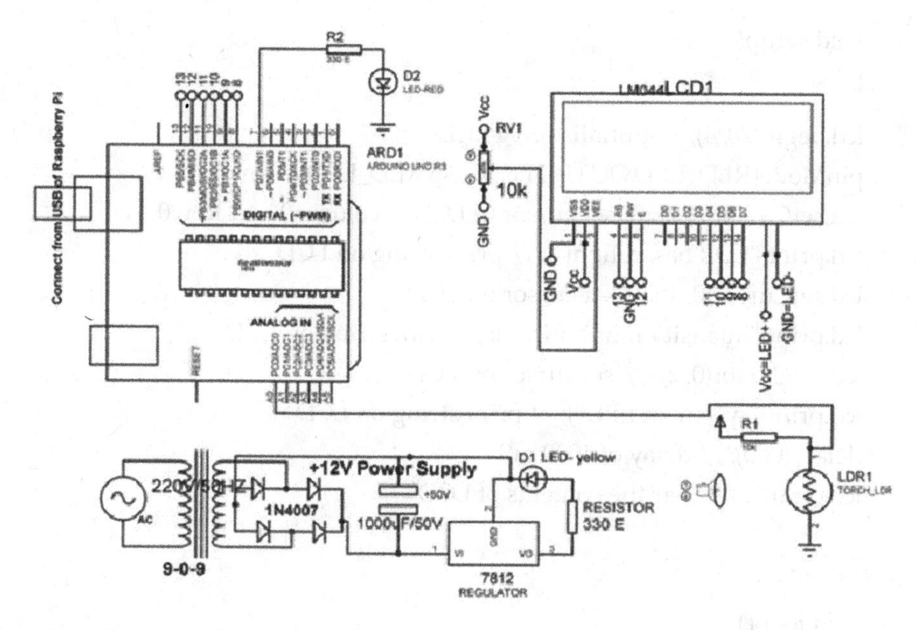

FIGURE 6.5
Circuit diagram for LDR sensor interfacing with Arduino.

LCD connection

- Connect Arduino digital pin 13 to RS pin(4) of LCD.
- Connect Arduino digital pin GND to RW pin(5) of LCD.
- Connect Arduino digital pin 12 to E pin(6) of LCD.
- Connect Arduino digital pin 11 to D4 pin(11) of LCD.
- Connect Arduino digital pin 10 to D5 pin(12) of LCD.
- Connect Arduino digital pin 9 to D6 pin(13) of LCD.
- Connect Arduino digital pin 8 to D7 pin(14) of LCD.

6.3.2 Sketch

```
#include <LiquidCrystal.h> // include library of LCD
LiquidCrystallcd(13, 12, 11, 10, 9, 8); // attach LCD pin RS,E,D4,D5,D6,D7 to the given pins
intLDR_sensor_Pin = A0; // select the input pin for the potentiometer
intLDR_sensor_ADC_Value = 0; // variable to store the value coming from the sensor
int RED_LED=7; // assign pin 7 to RED_LED
```

```
void setup()
{
lcd.begin(20, 4); // Initialise 20*4 LCD
pinMode(RED_LED,OUTPUT); // use RED_LED as an output
lcd.setCursor(0, 0); // set cursor of LCD at column0 and Row0
lcd.print("LDR based light"); // print string on LCD
lcd.setCursor(0, 1); // set cursor on LCD
lcd.print("intensity monitoring"); // print string on LCD
lcd.setCursor(0, 2); // set cursor on LCD
lcd.print("system at LPU"); // print string on LCD
delay(1000); // delay of 1000 mS
lcd.clear(); // clear the contents of LCD
}

void loop()
{
LDR_sensor_ADC_Value = analogRead(LDR_sensor_Pin); // read the
   value from the sensor
lcd.setCursor(0,2); // set cursor on LCD
lcd.print("ADC LEVEL+LDR:"); // print string on LCD
lcd.setCursor(17,2); // set cursor on LCD
lcd.print(LDR_sensor_ADC_Value); // // print value on LCD
if(LDR_sensor_ADC_Value>=100)
  {
digitalWrite(RED_LED,HIGH); // make pin7 to HIGH
delay(20); // delay of 20 mS
  }
else
  {
digitalWrite(RED_LED,LOW); // make pin7 to HIGH
delay(20); // delay of 20 mS
  }
}
```

6.4 Play with Actuators

An actuator is a component of a machine that is used for moving and controlling a mechanism or system. A DC motor, stepper motor, and servo motor are commonly used actuators in the systems.

6.4.1 DC Motor

The DC-geared motors with 100 rpm 12 V are generally used for robotics applications. They are very easy to use. They have nuts and threads on the shafts to easily connect and internally threaded shafts for easy connection to the wheel. Figure 6.6 shows the block diagram to interface the DC motor with Arduino. It is comprised of an Arduino Uno, a power supply, a liquid crystal display, a motor driver (L293D), and two DC motors.

6.4.1.1 Circuit Diagram

Connect the components as shown in Figure 6.7 to check the workings of a DC motor. Upload the program described in Section 6.4.1.2 and check the workings.

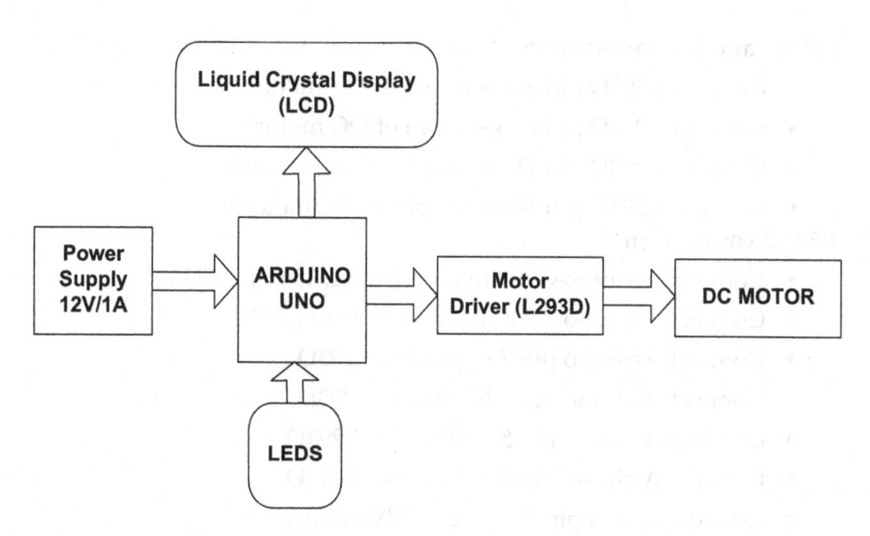

FIGURE 6.6
Block diagram of DC motor interfacing with Arduino.

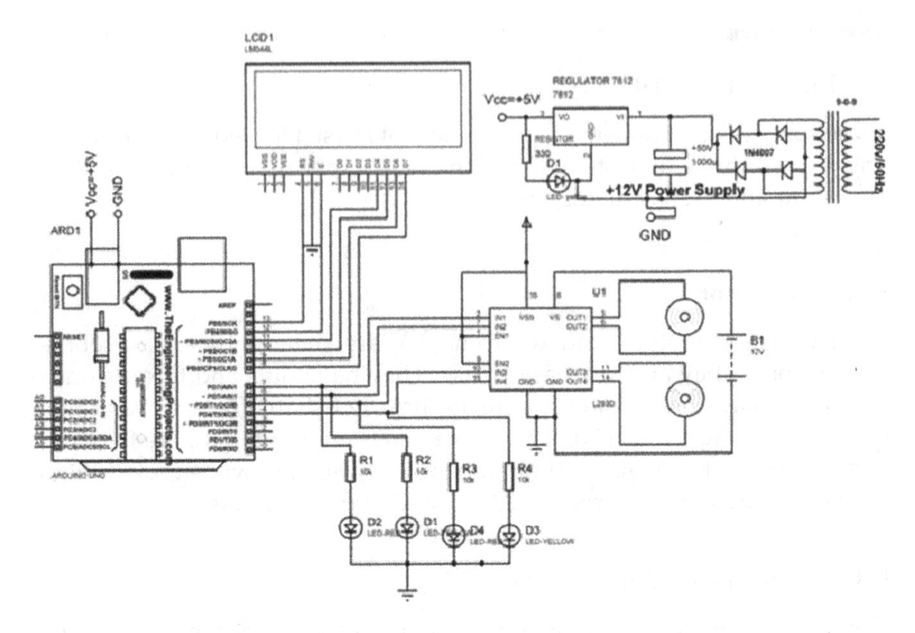

FIGURE 6.7
Circuit diagram of DC motor interfacing with Arduino.

L293D and DC motor connection

- Connect L293D pin 3 to +ve pin of DC motor1.
- Connect L293D pin 6 to –ve pin of DC motor1.
- Connect L293D pin 11 to +ve pin of DC motor2.
- Connect L293D pin 14 to +ve pin of DC motor2.

L293D connection

- Connect Arduino GND to pins 4, 5, 12, 13 of L293D.
- Connect Arduino +5 V to pins 1, 9, 16 of L293D.
- Connect Arduino pin 7 to pin 2 of L293D.
- Connect Arduino pin 6 to pin 7 of L293D.
- Connect Arduino pin 5 to pin 10 of L293D.
- Connect Arduino pin 4 to pin 15 of L293D.
- Connect L293D pin 8 to +ve of 12V battery

LED connection

- Connect Arduino pin 7 to anode of LED1.
- Connect Arduino pin 6 to anode of LED2.
- Connect Arduino pin 5 to anode of LED3.

- Connect Arduino pin 4 to anode of LED4.
- Connect cathode of all LEDs to ground.

LCD connection

- Connect Arduino digital pin 13 to RS pin(4) of LCD.
- Connect Arduino digital pin GND to RW pin(5) of LCD.
- Connect Arduino digital pin 12 to E pin(6) of LCD.
- Connect Arduino digital pin 11 to D4 pin(11) of LCD.
- Connect Arduino digital pin 10 to D5 pin(12) of LCD.
- Connect Arduino digital pin 9 to D6 pin(13) of LCD.
- Connect Arduino digital pin 8 to D7 pin(14) of LCD.

6.4.1.2 Sketch

```
#include <LiquidCrystal.h> // include library of LCD
LiquidCrystallcd(13, 12, 11, 10,9, 8); // attach LCD pin RS, E, D4, D5, D6,
   D7 to the given pins
int MPIN1= 7; // assign pin 7 as MPIN1
int MPIN2= 6; // assign pin 6 as MPIN2
int MPIN3= 5; // assign pin 5 as MPIN3
int MPIN4= 4; // assign pin 4 as MPIN4

void setup()
{
pinMode(MPIN1, OUTPUT); // make MPIN1 as an output
pinMode(MPIN2, OUTPUT); // make MPIN2 as an output
pinMode(MPIN3, OUTPUT); // make MPIN3 as an output
pinMode(MPIN4, OUTPUT); // make MPIN4 as an output
lcd.begin(20,4); // initialise LCD
lcd.setCursor(0, 0); // set cursor on LCD
lcd.print("DC Motor direction"); // print string on LCD
lcd.setCursor(0, 1); // set cursor on LCD
lcd.print("control system..."); // print string on LCD
delay(1000); // delay of 1000 mS
lcd.clear(); // clear the contents of LCD
}
void loop() // infinite loop
{
digitalWrite(MPIN1, HIGH); // make MPIN1 to HIGH
```

```
digitalWrite(MPIN2, LOW); // make MPIN2 to LOW
digitalWrite(MPIN3, HIGH); // make MPIN3 to HIGH
digitalWrite(MPIN4, LOW); // make MPIN4 to LOW
lcd.setCursor(0, 2); // set cursor on LCD
lcd.print("CLOCKWISE"); // print string on LCD
delay(2000); // delay of 2 sec
lcd.clear(); // clear the contents of LCD
digitalWrite(MPIN1, LOW); //make MPIN1 to LOW
digitalWrite(MPIN2, HIGH); //make MPIN2 to HIGH
digitalWrite(MPIN3, LOW); //make MPIN3 to LOW
digitalWrite(MPIN4, HIGH); //make MPIN4 to HIGH
lcd.setCursor(0, 2); // set cursor on LCD
lcd.print("ANTI-CLOCKWISE"); // print string on LCD
delay(2000); // delay of 2 Sec
lcd.clear(); // clear the contents of LCD
digitalWrite(MPIN1, LOW); // make MPIN1 to LOW
digitalWrite(MPIN2, LOW); //make MPIN2 to LOW
digitalWrite(MPIN3, HIGH); //make MPIN3 to HIGH
digitalWrite(MPIN4, LOW); //make MPIN4 to LOW
lcd.setCursor(0, 2); // set cursor on LCD
lcd.print("LEFT"); // print string on LCD
delay(2000); // delay of 2 Sec
lcd.clear(); // clear the contents of LCD
digitalWrite(MPIN1, HIGH); //make MPIN1 to HIGH
digitalWrite(MPIN2, LOW); //make MPIN2 to LOW
digitalWrite(MPIN3, LOW); //make MPIN3 to LOW
digitalWrite(MPIN4, LOW); //make MPIN4 to LOW
lcd.setCursor(0, 2); // set cursor on LCD
lcd.print("RIGHT"); // print string on LCD
delay(2000); // delay of 2 Sec
lcd.clear(); // clear the contents of LCD
  }
```

6.4.2 Servo Motor

A servo motor is a rotary actuator used for precise control of the angular position. It is comprised of a motor coupled with a sensor for

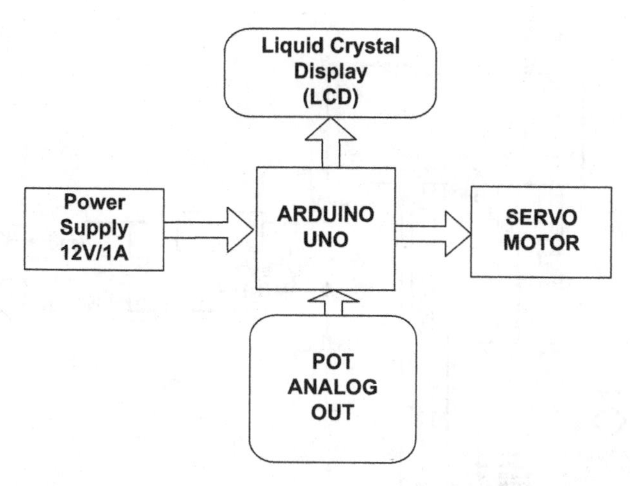

FIGURE 6.8
Block diagram to interface servo motor with Arduino.

position feedback. It also requires a servo drive. The drive uses the feedback sensor to control the rotary position of the motor precisely. This is called a closed-loop operation. The high torque standard servo motor with metal gears and 360° rotation can provide 11 kg/cm at 4.8 V, 13.5 kg/cm at 6 V, and 16 kg/cm at 7.2 V. Figure 6.8 shows the block diagram to interface the servo motor with Arduino. It is comprised of an Arduino Uno, a power supply, a liquid crystal display, a potentiometer (POT), and a servo motor. The system is designed to control the angle of the servo motor with the potentiometer.

6.4.2.1 Circuit Diagram

Connect the components as shown in Figure 6.9 to check the workings of a servo motor. Upload the program described in Section 6.4.2.2 and check the workings.

Servo connection

- Connect Arduino GND to GND pin of servo motor.
- Connect Arduino +5 V to "+" terminal of servo motor.
- Connect Arduino pin(3) to PWM pin of servo motor.

POT connection

- Connect Arduino GND to GND pin of POT.
- Connect Arduino +5 V to "+" terminal of POT.
- Connect Arduino A0 pin to data out pin of POT.

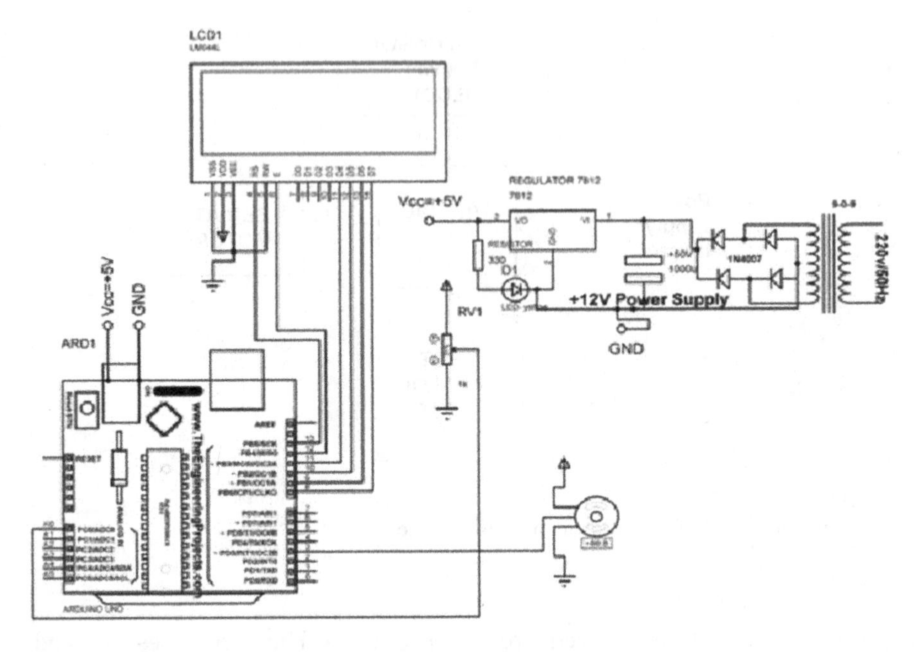

FIGURE 6.9
Circuit diagram to interface servo motor with Arduino.

LCD connection

- Connect Arduino digital pin (13) to RS pin(4) of LCD.
- Connect Arduino digital pin (GND) to RW pin(5) of LCD.
- Connect Arduino digital pin (12) to E pin(6) of LCD.
- Connect Arduino digital pin (11) to D4 pin(11) of LCD.
- Connect Arduino digital pin (10) to D5 pin(12) of LCD.
- Connect Arduino digital pin (9) to D6 pin(13) of LCD.
- Connect Arduino digital pin (8) to D7 pin(14) of LCD.

6.4.2.2 Sketch

```
#include <LiquidCrystal.h> // include library of LCD
LiquidCrystallcd(13, 12, 11, 10, 9, 8); // attach LCD pin RS,E,D4,D5,D6,D7
   to the given pins
Servo myservo; // create servo object to control a servo
int POT_PIN = A0; // analog pin used to connect the potentiometer
int POT_PIN_ADC_LEVEL; // variable to read the value from the
   analog pin
```

```
void setup()
{
myservo.attach(3); // attaches the servo on pin 9 to the servo object
lcd.begin(20,4); // initialise LCD
lcd.setCursor(0, 0); // set cursor on LCD
lcd.print("Servo ANALOG write "); // print string on LCD
lcd.setCursor(0, 1);// set cursor on LCD
lcd.print("system at LPU....");// print string on LCD
}
void loop()
{
POT_PIN_ADC_LEVEL = analogRead(POT_PIN); // reads POT value
   in the form of levels
POT_PIN_ADC_LEVEL = map(POT_PIN_ADC_LEVEL, 0, 1023, 0, 179);
   // map the value //between 0 to 180 degree for servo
myservo.write(POT_PIN_ADC_LEVEL); // sets the servo position
   according to the scaled value
lcd.setCursor(0, 2); // set cursor on LCD
lcd.print("ANGLE:"); // print string on LCD
lcd.print(POT_PIN_ADC_LEVEL); // print value on LCD
delay(15); // delay of 15 mSec
}
```

7

Python and Arduino with Pyfirmata

7.1 Python with Arduino

Arduino is an open-source platform to build hardware and software environments. Arduino provides limitless possibilities for tinkerers and electronics enthusiasts.

Raspberry Pi is a full-fledged computer that can do tasks like a desktop PC. It provides a platform for coding and designing electronic circuits, from creating a web server to a gaming console for retro gaming.

Arduino does not understand Python, so Firmata and Pyfirmata protocols are used to communicate through Raspberry Pi using Python. Pyfirmata is a protocol for Raspberry Pi to access Arduino. Firmata is protocol for Arduino to interface with Raspberry Pi with Python. The program will be written on Raspberry Pi in Python to access sensors connected to Arduino.

To install Firmata to Arduino, connect it to a USB socket of Raspberry Pi to communicate and power up Arduino. Next, install Firmata sketch to the Arduino in order for this open an Arduino IDE. Find the Firmata sketch in *File→Examples→Firmata→StandardFirmata* and upload it to the Arduino board. Once Firmata is installed, Arduino waits for communication from Raspberry Pi.

The next step is to install Pyfirmata to Raspberry Pi. For this, just run the following terminal commands on Raspberry Pi:

$ sudo apt-get install git

$ sudo git clone https://github.com/tino/pyFirmata.git

$ cdpyFirmata

$ sudo python setup.py install'

7.2 Controlling Arduino with Python

A "USB standard A" connector is used to connect Arduino with the Raspberry Pi. Now check for the USB address of Arduino by running *"ls -lrt /dev/tty*."* On my Raspberry Pi, it is listed as */dev/ttyUSB0* (Remember this value for later).

Import the Arduino and util classes from the Pyfirmata module to control an Arduino from a Python script on the Raspberry Pi. After this, create an object that was found in the previous step with the help of a USB address.

```
>>>from pyfirmata import Arduino, util
>>>board = Arduino('/dev/ttyUSB0')
```

7.3 Play with LED

The objective of the project is to control the Arduino digital output through Raspberry Pi with Python. To build this project, connect an LED to a digital pin of Arduino and write a short Python program to make it blink. Figure 7.1 shows the circuit diagram for the interfacing of an LED. The system is comprised of a Raspberry Pi3, an Arduino Uno, a power supply, and two LEDs connected to Pin6 and Pin7 of Arduino. The program is written to make LEDs blink after some time delay.

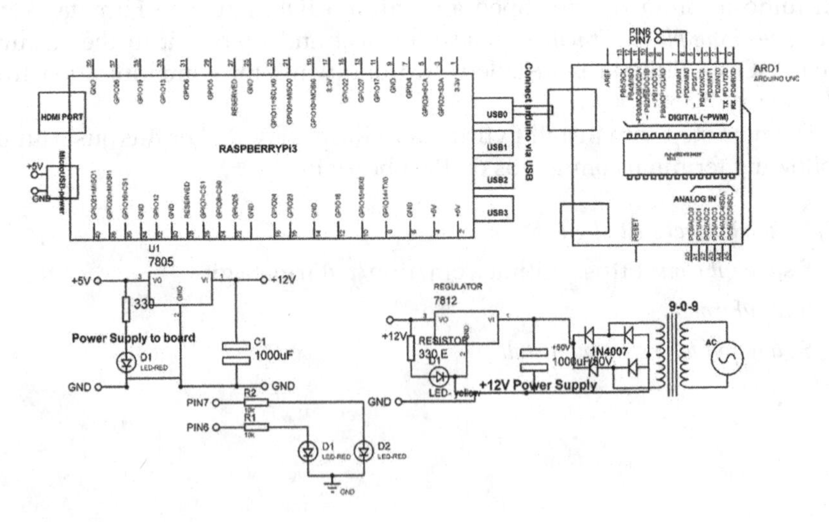

FIGURE 7.1
Circuit diagram for the interfacing of LED.

7.3.1 Recipe

```
import pyfirmata # import lib of pyfirmata
import time as wait    # import lib of pyfirmata
board = pyfirmata.Arduino('/dev/ttyUSB0')# define COM port of Arduino
red_pin = board.get_pin('d:7:o')# assign digital pin 7 as an output
green_pin = board.get_pin('d:6:o')# assign digital pin 6 as an output

whileTrue:    # infinite loop
        red_pin.write(1)# write '1' on pin 7
        green_pin.write(1)# write '1' on pin 6
        wait.sleep(0.5)# delay of 0.5 Sec
        red_pin.write(0)#write '0' on pin 7
        green_pin.write(0)# write '0' on pin 6
        wait.sleep(0.5)# delay of 0.5 Sec
```

7.4 Reading an Arduino Digital Input with Pyfirmata

The objective is to read the digital pins of Arduino on a Raspberry Pi by Python. Pyfirmata is used to read a digital input on Arduino. The components required for the recipe are Arduino Uno, 1 kΩ resistor, and a push switch or button (as digital sensor). A switch can be connected in two arrangements: pull down and pull up. The output of a digital pin of Arduino is normally "LOW," and digital sensors are available in two configurations for output: active "LOW" and active "HIGH." The pull-down arrangement is used where digital pin is normally "LOW," and on reading the sensor it gets "HIGH." This is used for the sensor that has the output as active "HIGH" on occurrence of an event; otherwise, the output is "LOW." The pull-up arrangement is for the sensor that has a normal output as active "HIGH," and on occurrence of an event it gets "LOW." In this arrangement, the digital pin needs to be activated as "HIGH" in the program so that it can read the sensor. Figure 7.2 shows circuit diagram for pull down, and Figure 7.3 shows circuit diagram for pull up.

As discussed in Section II of this book, the Pyfirmata protocol is used to read the input pin of Arduino by Raspberry Pi. It uses the concept of an iterator to monitor the Arduino pin. The iterator manages the reading of the switch using the following commands:

```
it = pyfirmata.util. Iterator(board)
it.start()
```

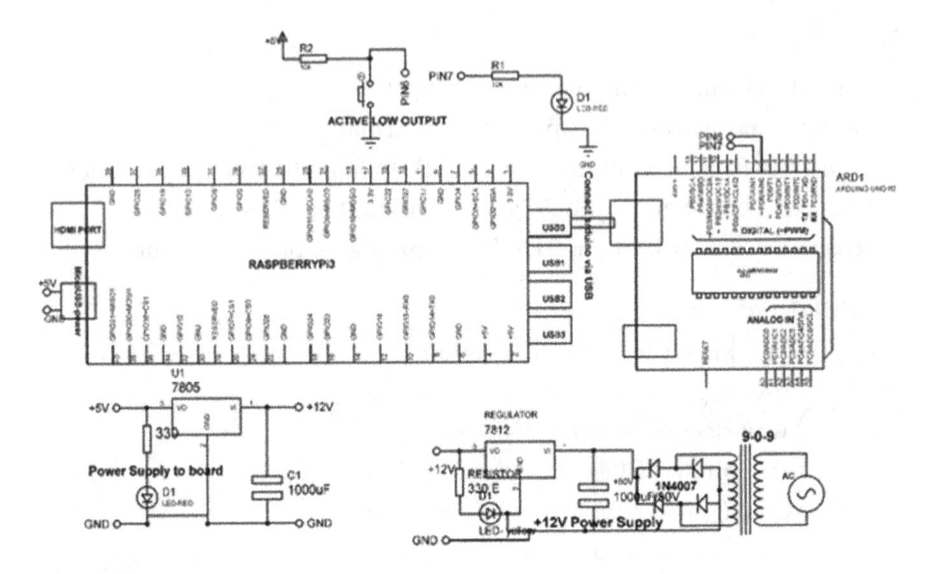

FIGURE 7.2
Pull-down arrangement for reading button.

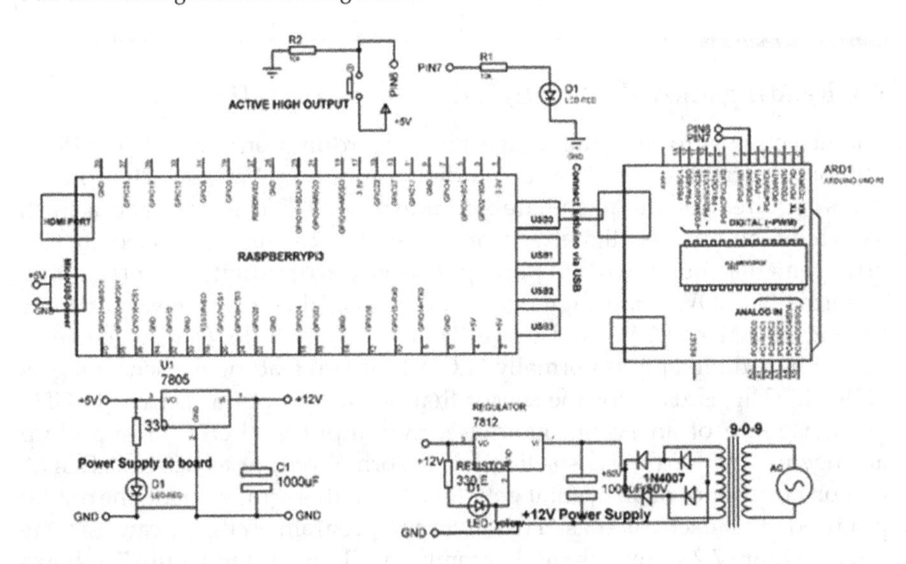

FIGURE 7.3
Pull-up arrangement for reading button.

After this enable the pin by using following command.

switch_pin.enable_reporting()

The iterator function can't be stopped, so when Ctrl+Z is pressed to exit the window, it will not exist.

To stop this function, simply disconnect Arduino from Raspberry Pi or open another terminal window and use the kill command:

$ sudokillall python

7.4.1 Recipe to Read Pull-Down Arrangement

```
import pyfirmata # import library of pyfirmata
import time as wait # import library of time
board = pyfirmata.Arduino('/dev/ttyUSB0') # define COM port of Arduino
button_pin = board.get_pin('d:6:i') # define pin 6 as an input
led_pin = board.get_pin('d:7:o') # define pin7 as an output
it = pyfirmata.util.Iterator(board) # use iterator
it.start() # start iterator
button_pin.enable_reporting() # enable input
while True: # infinite loop
        switch_state = switch_pin.read() # read input from pin 6
        ifswitch_state == False: # check condition
                print('Button Pressed') # print string on Pi terminal
                led_pin.write(1) # write '1' on pin 7
                wait.sleep(0.2) # delay of 0.2 Sec
else
                print('Button not Pressed') # print string on Pi terminal
                led_pin.write(0) # write '0' on pin 7
                wait.sleep(0.2) # delay of 0.2 Sec
```

7.4.2 Recipe to Read Pull-Up Arrangement

```
import pyfirmata # import library of pyfirmata
import time as wait # import library of time
board = pyfirmata.Arduino('/dev/ttyUSB0') # define COM port of Arduino
button_pin = board.get_pin('d:6:i') # assign pin 6 as digital input
led_pin = board.get_pin('d:7:o') # assign pin 7 as digital output
it = pyfirmata.util.Iterator(board) # use iterator
it.start() # start iterator
button_pin.enable_reporting() # enable pin
while True: # infinite loop
        switch_state = switch_pin.read() # read digital pin
        ifswitch_state == True: # check condition
                print('Button Pressed') # print string on Pi terminal
```

```
            led_pin.write(1) # make pin 7 to '1'
            wait.sleep(0.2) # delay of 0.2 Sec
    else
            print('Button not Pressed') # print string on Pi terminal
            led_pin.write(0) # make pin 7 to '1'
            wait.sleep(0.2) # delay of 0.2 Sec
```

7.5 Reading the Flame Sensor with Pyfirmata

The objective is to read the flame sensor as input with Python on a Raspberry Pi. A flame sensor can detect the infrared light with a wavelength ranging from 700 to 1000 nm. The far-infrared flame probe converts the detected light in the form of infrared light into current. It has a working voltage of 3.3 to 5.2 V DC, with a digital output to indicate the presence of a signal. An onboard LM393 comparator is used for condition sensing. Connect the components as shown in Figure 7.4 and check the workings by uploading the recipe described in Section 7.5.1.

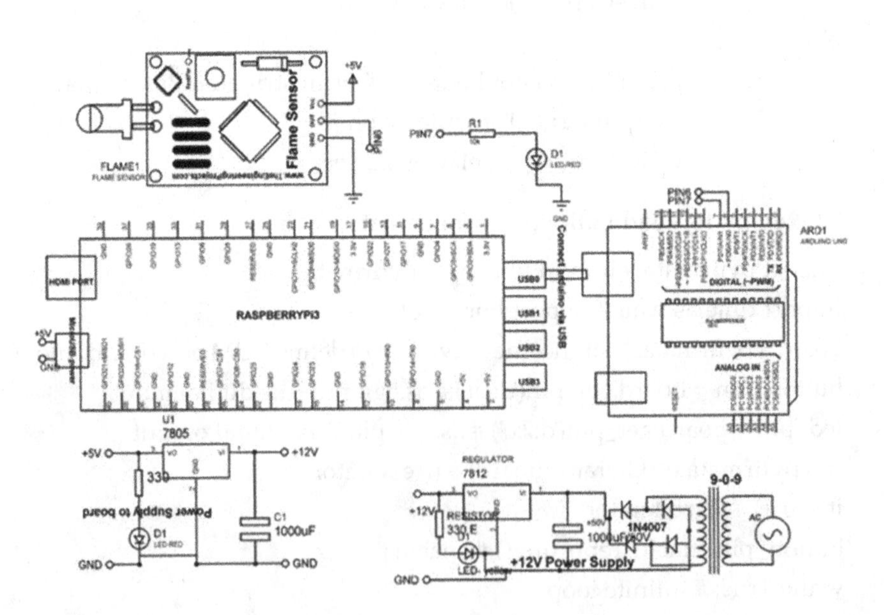

FIGURE 7.4
Circuit diagram for flame sensor interfacing.

7.5.1 Program for Reading Active "Low" Flame Sensor

```
import pyfirmata # import library of pyfirmata
import time as wait # import library of time
board = pyfirmata.Arduino('/dev/ttyUSB0') # define COM port of Arduino
flame_pin = board.get_pin('d:6:i') # assign pin 6 as digital input
indicator_pin = board.get_pin('d:7:o') # assign pin 7 as digital output
it = pyfirmata.util.Iterator(board) # use iterator
it.start() # start iterator
flame_pin.enable_reporting() # enable input
while True: # infinite loop
        flame_state = flame_pin.read() # read digital input
        ifflame_state == False: # check condition
                print('No Obstacle') # print string on Pi Terminal
                indicator_pin.write(1) # write '1'on pin7
                wait.sleep(0.2) # sleep for 0.2 sec
    else:
                print("Obstacle Found")) # print string on Pi Terminal
                indicator_pin.write(0) # write '0' on pin7
                wait.sleep(0.2) # sleep for 0.2 sec
```

7.6 Reading an Analog Input with Pyfirmata

A potentiometer is used to demonstrate the workings of the analog sensor with Pyfirmata. It is connected to pin A0 of Arduino (Figure 7.5). When the pin gets configured as an analog input pin in a program, it starts sending the input values to the serial port. If the data can't be managed properly, the data starts getting buffered at the serial port and quickly overflows; this situation can be handled with the program.

The Pyfirmata library has the reporting and iterator methods to overcome this situation. The *enable_reporting()* method is used to set the input pin to start reporting. This method is applied before performing a reading operation on the pin:

board.analog[3].enable_reporting()

Once the reading operation is done, the pin is set to disable reporting:

board.analog[3].disable_reporting()

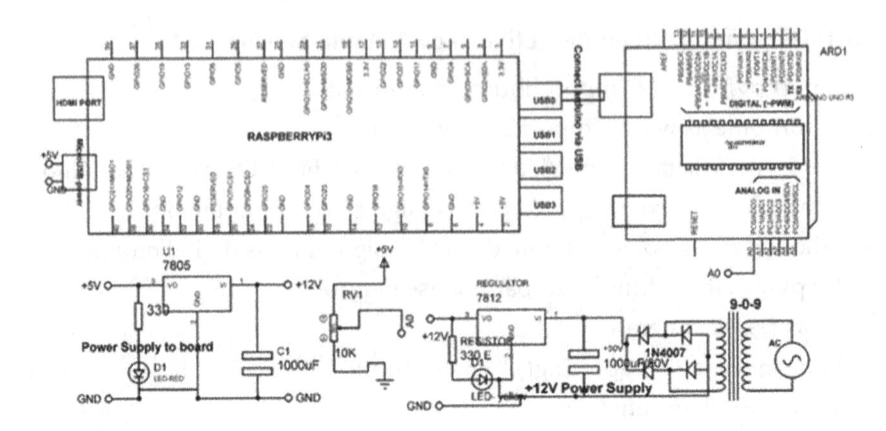

FIGURE 7.5
Circuit diagram for potentiometer interfacing.

To read the analog pin, *iteratorthread* is used in the main loop.

This class is defined in the util module of the Pyfirmata package and is imported before it getting utilized in the code:

```
from pyfirmata import Arduino, util
# Setting up the Arduino board
port = 'COM3'
board = Arduino(port)
sleep(5)
it = util.Iterator(board) # Start Iterator to avoid serial overflow
it.start()
board.analog[3].enable_reporting()
```

7.6.1 Recipe

```
import pyfirmata # import library of pyfirmata
import time as wait # import library of time
board = pyfirmata.Arduino('/dev/ttyUSB0') # define COM port of
   Arduino
POT_pin = board.get_pin('a:0:i') # assign A0 pin as an input
it = pyfirmata.util.Iterator(board) # use iterator
it.start() # start iterator
POT_pin.enable_reporting() # enable pin
while True: # infinite loop
POT_reading = POT_pin.read() # read analog pin
```

```
if POT_reading != None: # check condition
        POT_voltage = POT_reading * 5.0 # convert levels to
          voltage
        print("POT_reading=%f\t POT_voltage=%f"% (POT_
          reading, POT_voltage))
        # printvalues on Pi terminal
        wait.sleep(1) # sleep for 1 sec
else:
        print("No reading Obtained") # print string on Pi terminal
        wait.sleep(1)# sleep for 1 sec
```

7.7 Reading the Temperature Sensor with Pyfirmata

The LM35 series of temperature sensors has an output voltage linearly proportional to the Centigrade temperature. The LM35 device does not require any calibration or trimming to provide the accuracy of ±¼°C at room temperature and has a sensing range of −55°C to 150°C. The LM35 device draws a 60-μA current from the supply. The LM35 series devices are available in hermetic TO transistor packages, while the LM35C, LM35CA, and LM35D are available in the plastic TO-92 transistor packages. Figure 7.6 shows the circuit diagram of the LM35 interfacing. The output of LM35 is connected to the A0 pin of Arduino.

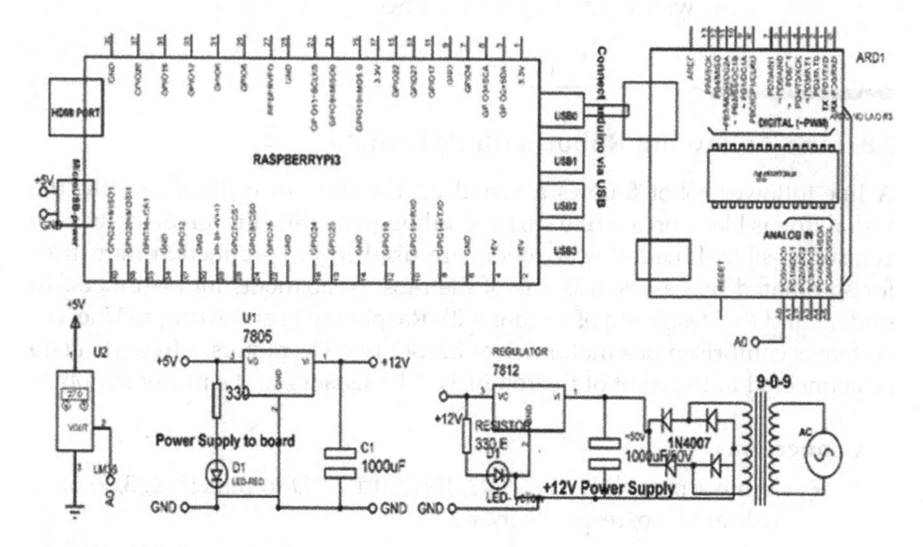

FIGURE 7.6
Circuit diagram of LM35 interfacing.

7.7.1 Recipe

```
IMPORT pyfirmata # import library of pyfirmata
import time as wait # import library of time
board = pyfirmata.Arduino('/dev/ttyUSB0') # define COM port of
   Arduino
POT_pin = board.get_pin('a:0:i') # assign A0 pin as an input
it = pyfirmata.util.Iterator(board) # use iterator
it.start() # start iterator
POT_pin.enable_reporting() # enable pin
while True:
reading = switch_pin.read() # read analog input
if reading != None: # check condition
            voltage = reading * 5.0 # convert level into voltage
            temp = (voltage*1000)/10 # convert voltage into
               temperature
            print('Reading=%f\t Voltage=%f\tTemperature=%f'%
               (reading,voltage,temp))
            # print value on Pi Terminal
            wait.sleep(1) # sleep for 1 Sec
else:
            print("No reading Obtained") # print string on Pi Terminal
            wait.sleep(1)# sleep for 1 Sec
```

7.8 Line-Following Robot with Pyfirmata

A line-follower robot follows a visual on the floor or ceiling. Usually, the visual line is black on a white surface, although a white line on black surface is also possible. Line-follower robots are used in the production industries for automated processes. It is one of the most basic robots for beginners. To understand the designing of a robot with Raspberry Pi and Arduino Uno, the system is comprised of a motor driver L293D, two DC motors, a free wheel (to be connected in the front of the robot), two IR sensors, and a power supply.

Connections:

- Connect pins (IN1, IN2, IN3, IN4) of L293D to pins (5, 4, 3, 2) of Arduino Uno, respectively.
- Connect a DC motor (M1) between pins (OUT1 and OUT2) of L293D.

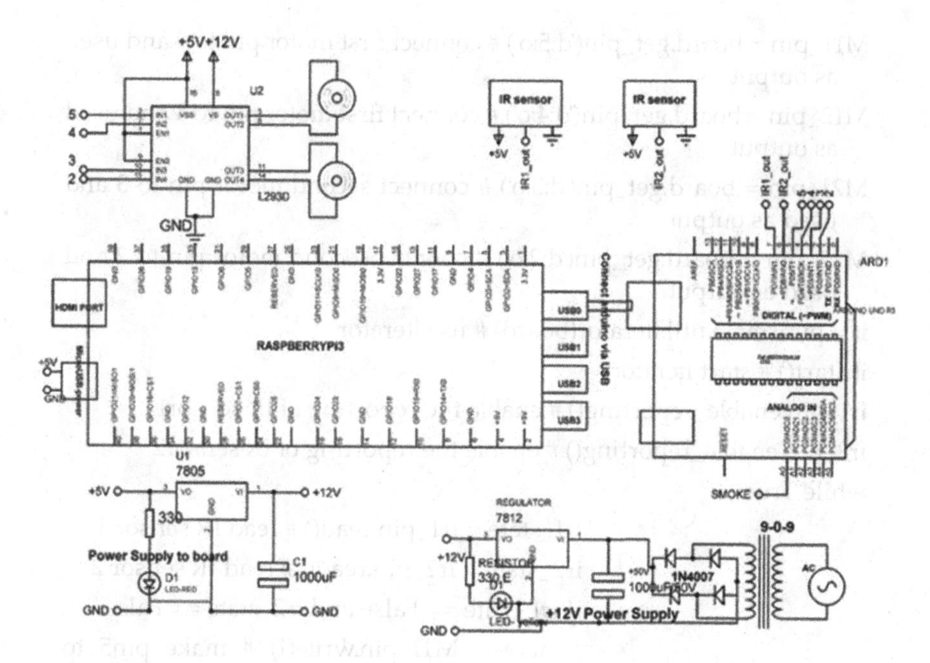

FIGURE 7.7
Circuit diagram for line-following robot.

- Connect other DC motor (M2) between pins (OUT3 and OUT4) of L293D.
- Connect pins (Vcc and ground) of IR1 and IR2 to +5 VDC and ground, respectively.
- Connect pin (OUT) of IR1 to pin (7) of Arduino Uno.
- Connect pin (OUT) of IR2 to pin (6) of Arduino Uno.
- Connect Arduino Uni to Raspberry Pi through a USB.

Figure 7.7 shows the circuit diagram for a line-following robot.

7.8.1 Recipe

```
import pyfirmata
import time as wait
board = pyfirmata.Arduino('/dev/ttyUSB10')
ir1_pin = board.get_pin('d:7:i') # connect IR sensor1 to pin 7 and used
   as input
ir2_pin = board.get_pin('d:6:i') # connect IR sensor2 to pin 6 and used
   as input
```

```
M11_pin = board.get_pin('d:5:o') # connect first motor pin to 5 and used
    as output
M12_pin = board.get_pin('d:4:o') # connect first motor pin to 4 and used
    as output
M21_pin = board.get_pin('d:3:o') # connect second motor pin to 3 and
    used as output
M22_pin = board.get_pin('d:2:o') # connect second motor pin to 2 and
    used as output
it = pyfirmata.util.Iterator(board) # use iterator
it.start() # start iterator
ir1_pin.enable_reporting() # enable the reporting of IR sensor1
ir2_pin.enable_reporting() # enable the reporting of IR sensor2
while True:
                    ir1_state = ir1_pin.read() # read IR sensor 1
                    ir2_state = ir2_pin.read() # read IR sensor 2
                    if ir1_state == False and ir2_state == False:
                            M11_pin.write(1) # make pin5 to
                              HIGH
                            M12_pin.write(0) # make pin4 to
                              LOW
                            M21_pin.write(1) # make pin3 to
                              HIGH
                      M22_pin.write(0) # make pin2 to LOW
                      print('forward') # print on terminal
                            wait.sleep(0.5) # delay of 500mSec
                    elif ir1_state == False and ir2_state == True:
                            M11_pin.write(1) # make pin5 to
                              HIGH
                            M12_pin.write(0) # make pin4 to
                              LOW
                            M21_pin.write(0) # make pin3 to
                              LOW
                            M22_pin.write(0) # make pin2 to
                              LOW
                      print('Left') # print on terminal
                      time.sleep(0.5) # delay of 500mSec
```

```
        elif ir1_state == True and ir2_state == False:
                M11_pin.write(0) # make pin5 to
                  LOW
                M12_pin.write(0) # make pin4 to
                  LOW
                M21_pin.write(1) # make pin3 to
                  HIGH
                M22_pin.write(0) # make pin2 to
                  LOW
                print('Right') # print on terminal
                time.sleep(0.5)# delay of 500mSec
        elif ir1_state == True and ir2_state == True:
                M11_pin.write(0) # make pin5 to
                  LOW
                M12_pin.write(0) # make pin4 to
                  LOW
                M21_pin.write(0) # make pin3 to
                  LOW
                M22_pin.write(0) # make pin2 to
                  LOW
          print('Stop') # print on terminal
          time.sleep(0.5) # delay of 500mSec
```

8

Python GUI with Tkinter and Arduino

8.1 Tkinter for GUI Design

The basic features for graphical user interface (GUI) libraries include the ease to install and minimal computations. Tkinter framework satisfies the basic requirements. It is also the default GUI library with the Python installations.

Tkinter interface is a cross-platform Python interface for the Tk GUI toolkit. Tkinter provides a layer on Python while Tk provides the graphical widgets. Tkinter is a cross-platform library that gets deployed as a part of the Python installation packages for major operating systems.

Tkinter is designed with minimal programming efforts for creating graphical applications. To test the version of the Tk toolkit, use the following commands on the Python prompt:

```
>>> import Tkinter
>>> Tkinter._test()
```

An image with the version information will be prompted (Figure 8.1).

If the window (Figure 8.1) is not visible, then reinstall Python.

The Tkinter interface supports various widgets to develop GUI. Table 8.1 describes the few widgets.

8.2 LED Blink

To understand the working of LED with Tkinter GUI, a simple circuit is designed. The system is comprised of a Raspberry Pi, an Arduino, an LED, and a power supply. An Arduino is connected to Raspberry Pi through a USB connector. The anode terminal of LED is connected to pin(7) of Arduino through a 10-K resistor, and the cathode terminal is connected to the ground. Figure 8.2 shows the circuit diagram to interface LED.

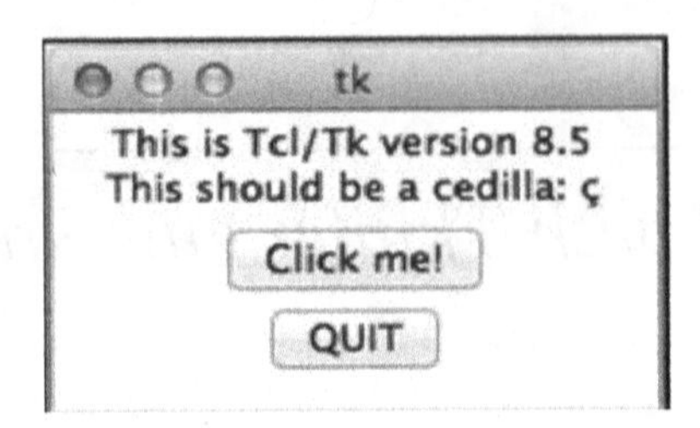

FIGURE 8.1
Tk version.

TABLE 8.1
The Tkinter Widgets to Develop GUI

Widget	Description
Tk()	Root widget need in every program
Label()	It shows an image or a text
Button()	Execute actions
Entry()	Text field to provide inputs
Scale()	It provides a numeric value by dragging the slider
Checkbox()	It enables toggling between two values

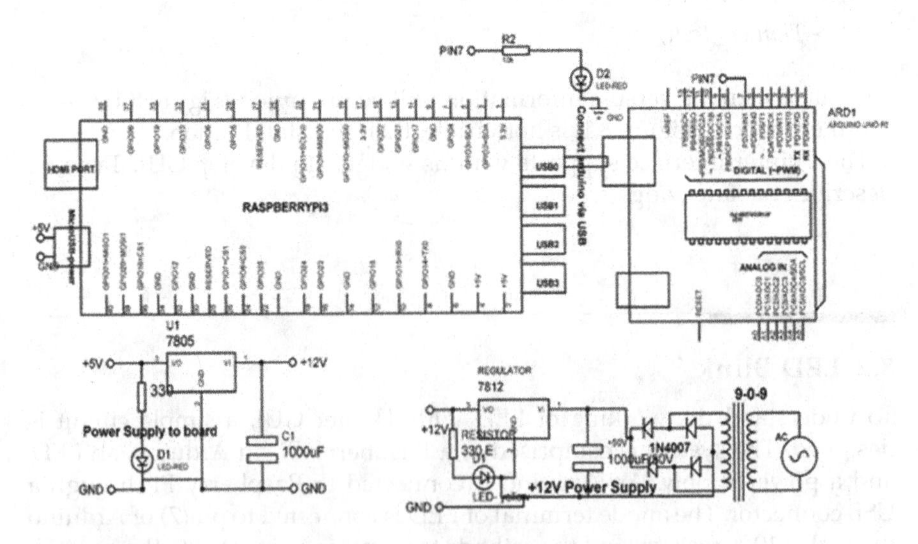

FIGURE 8.2
Circuit diagram to interface LED.

8.2.1 Recipe for LED Blinking with Fixed Time Delay

```
import Tkinter
import pyfirmata
import time as wait
# Associate port and board with pyfirmata
board = pyfirmata.Arduino('/dev/ttyUSB0')
wait.sleep(5) # delay of 5Sec
led_Pin = board.get_pin('d:7:o') # connect led to pin 7 and used as output
def call_LED_BLINK():
    button.config(state = Tkinter.DISABLED)
    led_Pin.write(1) # make led_Pin to HIGH
    print('LED at pin7 is ON') # print on terminal
wait.sleep(5) # delay of 5 sec
    print('LED at pin 7 is OFF') # print on terminal
    led_Pin.write(0) # make led_Pin to LOW
    button.config(state=Tkinter.ACTIVE)
# Initialize main windows with title and size
TOP = Tkinter.Tk()
TOP.title("Blink LED using button")
TOP.minsize(300,30)
# Create a button on main window and associate it with above method
button = Tkinter.Button(TOP, text="Press start to blink", command =
   call_LED_BLINK)
button.pack()
TOP.mainloop()
```

8.2.1.1 Tkinter GUI for LED Blinking with Fixed Delay

Run the program described in Section 8.2.1, and a GUI will appear (Figure 8.3).

After pressing the "Press start to blink" button on GUI, it will print "LED at pin7 is ON" for 5 sec and then print "LED at pin7 is OFF" (Figure 8.4).

8.2.2 Recipe for LED Blinking with Variable Delay

```
import Tkinter
import pyfirmata
```

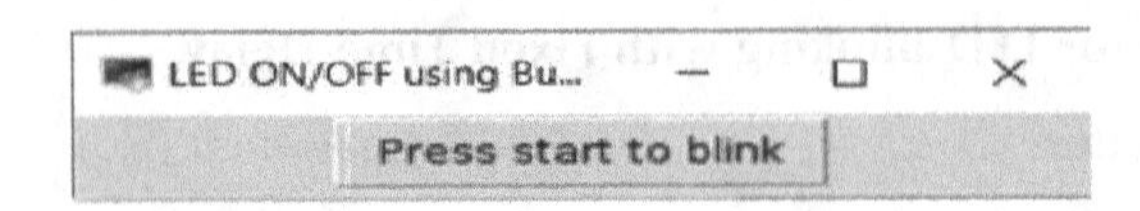

FIGURE 8.3
Tkinter GUI to control LED with fixed delay.

```
pi@raspberrypi:~ $ nano pyfir_led_tkinter_BOOK.py
pi@raspberrypi:~ $ python pyfir_led_tkinter_BOOK.py
LED at pin7 is ON
LED at pin 7 is OFF
```

FIGURE 8.4
LED ON/OFF.

```
import time as wait
board = pyfirmata.Arduino('/dev/ttyUSB0')
wait.sleep(5) # delay of 5 sec
led_Pin = board.get_pin('d:7:o') # connect led to pin 7 and used as output
def led_blink_variable_delay():
    # Value for delay is obtained from the Entry widget input
    time_Period = time_Period_Entry.get()
    time_Period = float(time_Period)
    button.config(state=Tkinter.DISABLED)
    led_Pin.write(1) # make led_Pin to HIGH
    print('pin7 connected led is ON') # print on terminal
wait.sleep(time_Period) # delay of 5 sec
    print('pin7 connected led is off') # print on terminal
    led_Pin.write(0) # make led_Pin to LOW
    button.config(state=Tkinter.ACTIVE)
TOP = Tkinter.Tk()
TOP.title("enter variable time")
time_Period_Entry = Tkinter.Entry(TOP, bd=6, width=28)
time_Period_Entry.pack()
time_Period_Entry.focus_set()
button = Tkinter.Button(TOP, text="start to blink", command=led_
  blink_variable_delay)
button.pack()
TOP.mainloop()
```

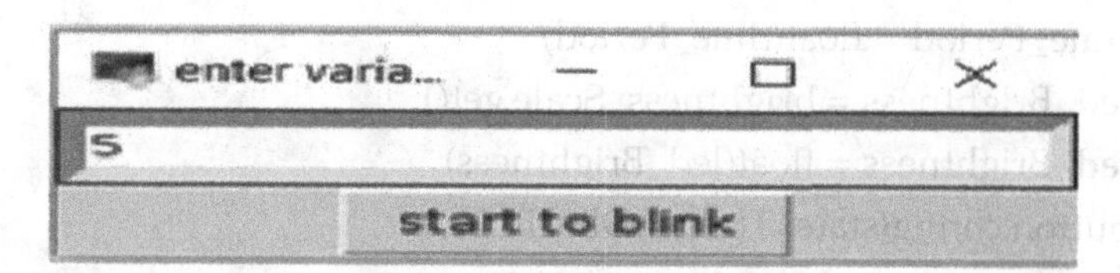

FIGURE 8.5
Tkinter GUI to control LED with variable delay.

```
pi@raspberrypi:~ $ nano pyfir_led_tkinter_variable_time_BOOK.py
pi@raspberrypi:~ $ python pyfir_led_tkinter_variable_time_BOOK.py
pin7 connected led is ON
pin7 connected led is off
```

FIGURE 8.6
Screenshot for LED ON/OFF with delay.

8.2.2.1 Tkinter GUI for LED Blinking with Variable Delay

Run the program described in Section 8.2.2, and a GUI will appear (Figure 8.5).
Write down the required delay in the blank slot, for example, take value 5. After pressing the "start to blink" button and delay of 5 sec on GUI, it will print "Pin7 connected is ON" for 5 sec and then print "Pin7 connected is OFF" (Figure 8.6).

8.3 LED Brightness Control

Figure 8.3 shows the circuit diagram for LED interfacing where LED is connected to pin (7) of Arduino. Pin(7) is also a PWM pin in Arduino Uno, so with the same circuit in Figure 8.3, LED brightness can be controlled with a different program.

8.3.1 Recipe

```
import Tkinter # add Tkinter library
import pyfirmata # add pyfirmata library
import time as wait # add time library
board = pyfirmata.Arduino('/dev/ttyUSB0')
wait.sleep(5) # delay of 5 Sec
led_Pin = board.get_pin('d:7:o') # connect led to pin 7 and used as output
def call_led_blink_pwm():
    time_Period = time_Period_Entry.get()
```

```
        time_Period = float(time_Period)
        led_Brightness = brightness_Scale.get()
        led_Brightness = float(led_Brightness)
        button.config(state=Tkinter.DISABLED)
        led_Pin.write(led_Brightness/100.0)
        print 'LED brightness control' # print on terminal
    wait.sleep(time_Period)
        led_Pin.write(0) # make led_Pin to LOW
        button.config(state=Tkinter.ACTIVE)

    TOP = Tkinter.Tk()
    time_Period_Entry = Tkinter.Entry(TOP, bd=7, width=30)
    time_Period_Entry.pack()
    time_Period_Entry.focus_set()
    brightness_Scale = Tkinter.Scale(TOP, from_=0, to=100, orient=Tkinter.
      VERTICAL)
    brightness_Scale.pack()
    button = Tkinter.Button(TOP, text="Start", command =
      call_led_blink_pwm)
    button.pack()
    TOP.mainloop()
```

8.3.2 Tkinter GUI for LED Brightness Control

Run the program described in Section 8.3.1. Brightness can be controlled with a scale () widget. Once the LED is turned off after the time delay, the slider can be reset to another position to dynamically vary the value for the brightness. The slider can be placed horizontally instead of on a vertical scale. By clicking on "Start," it will display the message "LED brightness control" (Figures 8.7 and 8.8).

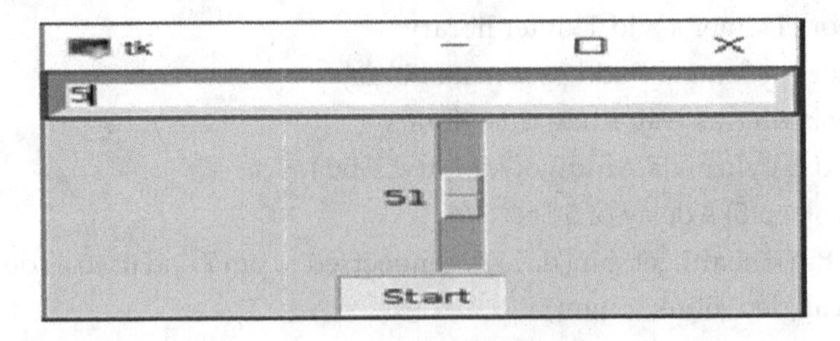

FIGURE 8.7
Tkinter GUI for LED brigtness control.

```
pi@raspberrypi:~ $ nano pyfir_led_pwm_tkinter_BOOK.py
pi@raspberrypi:~ $ python pyfir_led_pwm_tkinter_BOOK.py
LED brightness control
```

FIGURE 8.8
Screenshot for LED brightness control.

8.4 Selection from Multiple Options

When the user needs to select from multiple options from the given set of values, the complexity of the project increases. For example, when multiple numbers of LEDs are interfaced with the Arduino board, the user needs to select an LED or LEDs to turn it on. The Tkinter library provides an interface for a widget called Checkbutton(), which enables the selection process from the given options. The concept can be understood with a simple circuit of interfacing two LEDs with Arduino and Raspberry Pi using Pyfirmata.

The system is comprised of a Raspberry Pi, an Arduino Uno, two LEDs, and a power supply. Arduino is connected to Raspberry Pi through a USB. The anode terminals of LED1 and LED2 are connected to pin(6) and pin(7), respectively, through a 10-K resistor each, and cathode terminals of both LEDs are connected to ground (Figure 8.9).

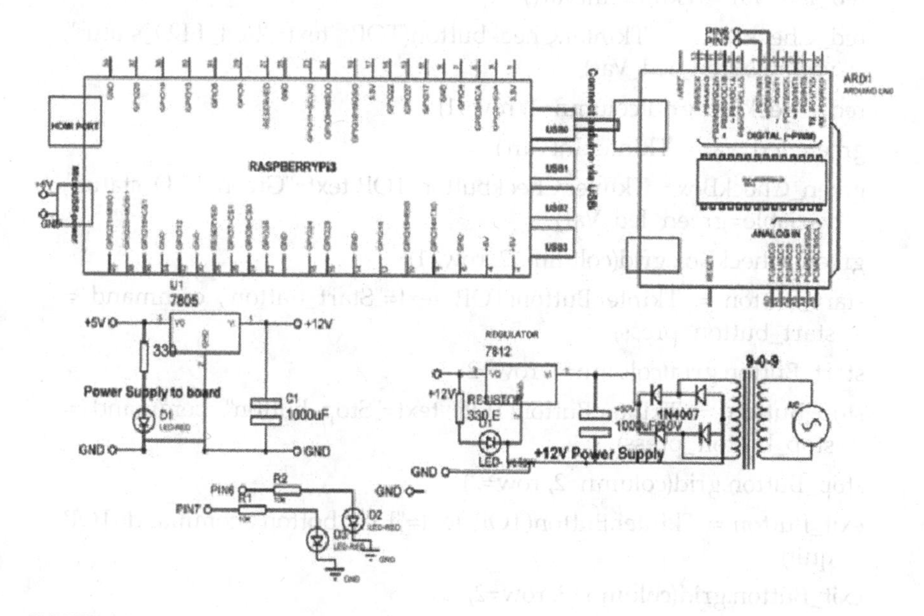

FIGURE 8.9
Circuit diagram to interface multiple LEDs.

8.4.1 Recipe

```
import Tkinter # add Tkinter library
import pyfirmata # add pyfirmata library
import time as wait # add time library
board = pyfirmata.Arduino('/dev/ttyUSB0')
wait.sleep(5) # delay of 5 Sec
red_led_pin = board.get_pin('d:7:o') # connect led to pin 7 and used as
  output
green_led_pin = board.get_pin('d:6:o') # connect led to pin 6 and used
  as output
def start_button_press():
    red_led_pin.write(red_led_Var.get())
    green_led_pin.write(green_led_Var.get())
    print 'start...' # print on terminal
def stop_button_press():
    red_led_pin.write(0) # make pin 7 to HIGH
    green_led_pin.write(0) # make pin 6 to HIGH
    print 'stop....' # print on terminal
TOP = Tkinter.Tk()
red_led_Var = Tkinter.IntVar()
red_CheckBox = Tkinter.Checkbutton(TOP, text="Red_LED_status",
  variable=red_led_Var)
red_CheckBox.grid(column=1, row=1)
green_led_Var = Tkinter.IntVar()
green_CheckBox = Tkinter.Checkbutton(TOP, text="Green_LED_status",
  variable=green_led_Var)
green_CheckBox.grid(column=2, row=1)
start_Button = Tkinter.Button(TOP, text="Start_button", command =
  start_button_press)
start_Button.grid(column=1, row=2)
stop_Button = Tkinter.Button(TOP, text="Stop_button", command =
  stop_button_press)
stop_Button.grid(column=2, row=2)
exit_Button = Tkinter.Button(TOP, text="Exit_button", command=TOP.
  quit)
exit_Button.grid(column=3, row=2)
TOP.mainloop()
```

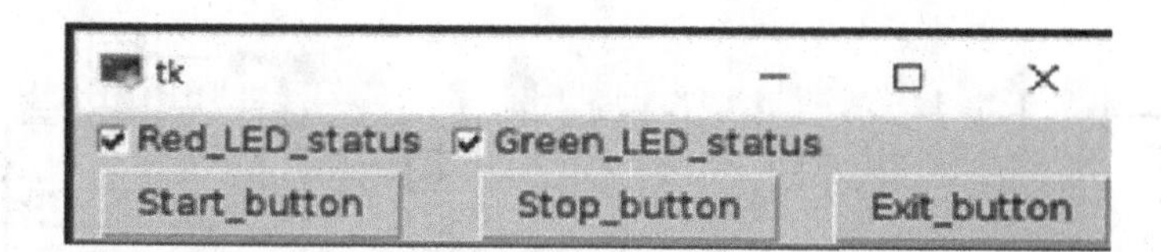

FIGURE 8.10
Tkinter GUI for multiple LEDs.

```
pi@raspberrypi:~ $ nano pyfir_two_led_tkinter_BOOK.py
pi@raspberrypi:~ $ python pyfir_two_led_tkinter_BOOK.py
start...
stop....
```

FIGURE 8.11
Screenshot for status window.

8.4.2 Tkinter GUI

To generate Tkinter GUI for multiple LEDs, run the program described in Section 8.4.1, and a window will appear (Figure 8.10). The GUI has two Checkbutton() widgets each for the red and green LED. The user can select the LEDs individually or together to make it start or stop (Figure 8.11).

8.5 Reading a PIR Sensor

A pyroelectric infrared (PIR) sensor is used to detect motion. A system is designed with a PIR sensor where two LEDs show the status of motion. If no motion is detected, the green LED will glow, and if motion is detected, an alert is generated with making the red LED "ON." The system is made ON/OFF with Tkinter GUI. The status of sensor prints on the Python prompt. The system is comprised of a Raspberry Pi, an Arduino, two LEDs, a PIR sensor, and a power supply. Arduino is connected to Raspberry Pi through a USB. The anode terminal of LED1 and LED2 are connected to pin (7) and pin(6) of Arduino, respectively. The pin (OUT) of the PIR sensor is connected to pin (8) of Arduino, and pin (Vcc) and pin (GND) are connected to +5 V DC and ground, respectively (Figure 8.12).

8.5.1 Recipe to Read PIR Sensor

```
# Define custom function to perform Blink action
def blink_LED(pin, message):
```

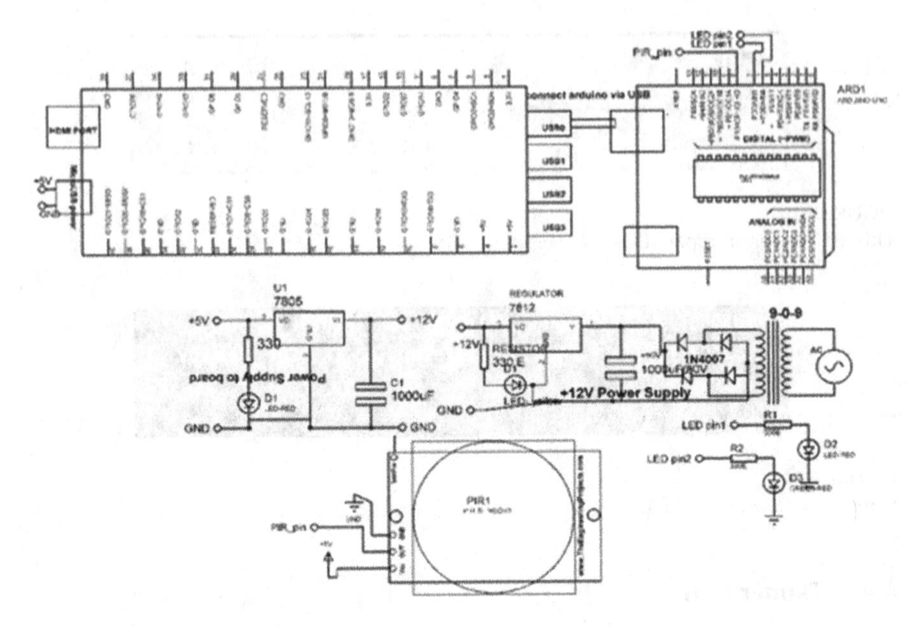

FIGURE 8.12
Circuit diagram for PIR sensor interfacing.

```
        Motion_Label.config(text=message)
        Motion_Label.update_idletasks()
        TOP.update()
        pin.write(1) # make pin to HIGH
    wait.sleep(1) # delay of 1 Sec
        pin.write(0) # make pin to HIGH
    wait.sleep(1) # delay of 1 Sec
    # Define the action associated with Start button press
    def press_start_button():
        while True:
            if FLAG.get():
                if PIR_pin.read() is True:
                    blink_LED(red_led_pin, "motion_status:Y")
                    print 'Motion detected' # print on terminal
                else:
                    blink_LED(green_led_pin, "motion_status:N")
                    print 'No motion' # print on terminal
```

```
        else:
            break
    board.exit()
    TOP.destroy()
def press_exit_button():
    FLAG.set(False)
import Tkinter # import Tkinter library
import pyfirmata # import pyfirmata library
import time as wait # import time library
board = pyfirmata.Arduino('/dev/ttyUSB0')
wait.sleep(5) # wait for 5Sec
PIR_pin = board.get_pin('d:8:i') # connect PIR sensor to pin8 and as
    input
red_led_pin = board.get_pin('d:7:o') # connect red LED to pin7 and as
    output
green_led_pin = board.get_pin('d:6:o') # connect Green led to pin6 and
    as output
# Using iterator thread to avoid buffer overflow
it = pyfirmata.util.Iterator(board)
it.start()                    # start iterator
PIR_pin.read() # read PIR sensor
# Initialize main windows with title and size
TOP = Tkinter.Tk()
TOP.title("PIR_sensor_for_motion")
# Create Label to for motion detection
Motion_Label = Tkinter.Label(TOP, text="Press Start..")
Motion_Label.grid(column=1, row=1)
# Create flag to work with indefinite while loop
FLAG = Tkinter.BooleanVar(TOP)
FLAG.set(True)
# Create Start button and associate it with onStartButtonPress method
Start_Button = Tkinter.Button(TOP , text="Start", command=press _start_
    button)
Start_Button.grid(column=1, row=2)
# Create Stop button and associate it with onStopButtonPress method
```

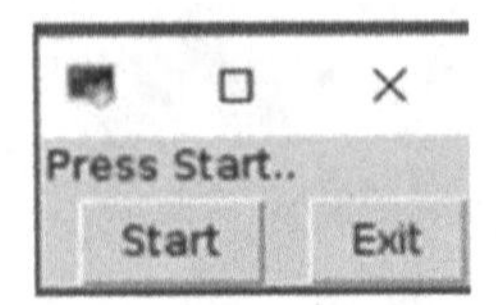

FIGURE 8.13
Tkinter GUI for motion detection.

```
pi@raspberrypi:~ $ nano pyfir_digital_in.tkinter_BOOK.py
pi@raspberrypi:~ $ python pyfir_digital_in.tkinter_BOOK.py
No motion
No motion
No motion
No motion
```

FIGURE 8.14
Status window for motion sensor.

```
Stop_Button = Tkinter.Button(TOP, text="Exit", command=press_exit_
   button)
Stop_Button.grid(column=2, row=2)
# Start and open the window
TOP.mainloop()
```

8.5.2 Tkinter GUI

To generate Tkinter GUI for motion status, run the program described in Section 8.5.1, and a window will appear (Figure 8.13). On pressing the "Start" button, it will give status of the motion sensor as "No motion" or "Motion detected" (Figure 8.14).

8.6 Reading an Analog Sensor

A potentiometer (POT) is analogous to an analog sensor. The change in output voltage can be observed on changing the resistance value by moving the knob. To create a Tkinter GUI for a POT, the system is comprised of a Raspberry Pi, an Arduino, a 10-K POT, and a power supply. Arduino is connected to Raspberry Pi through a USB. A POT has three terminals: one

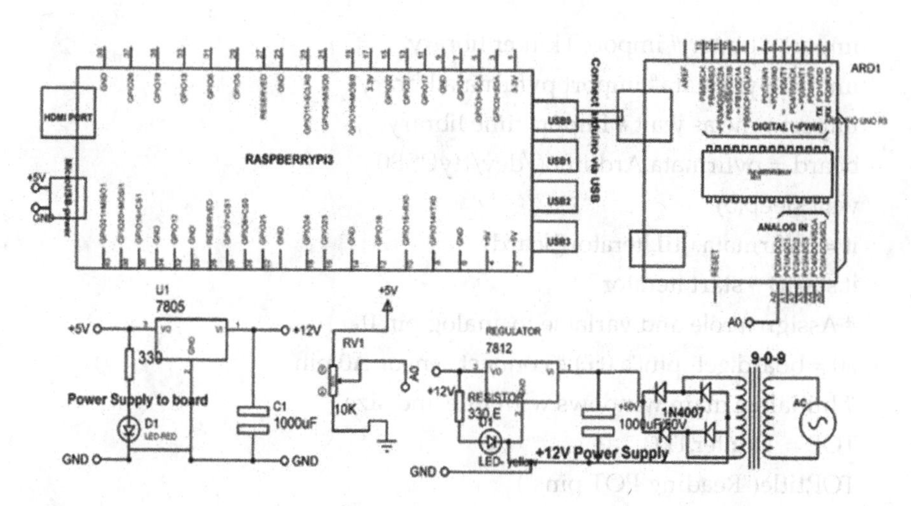

FIGURE 8.15
Circuit diagram for interfacing of POT.

terminal is connected to +5 V and other to ground. The wiper of POT is connected to pin (A0) of Arduino (Figure 8.15).

8.6.1 Recipe

```
# Define the action associated with Start button press
def start_button_press():
  while True:
    if FLAG.get():
      analog_Read_Label.config(text=str(a0.read()))
      analog_Read_Label.update_idletasks()
      TOP.update()
      print 'analog values of pot'
    else:
      break
  board.exit()
  TOP.destroy()

# Define the action associated with Exit button press
def exit_button_press():
  FLAG.set(False)
```

```
import Tkinter # import Tkinter library
import pyfirmata# import pyfirmata library
import time as wait # import time library
board = pyfirmata.Arduino('/dev/ttyUSB0')
wait.sleep(5)
it = pyfirmata.util.Iterator(board)
it.start() # start iterator
# Assign a role and variable to analog pin 0
a0 = board.get_pin('a:0:i') # connect sensor A0 pin
# Initialize main windows with title and size
TOP = Tkinter.Tk()
TOP.title("Reading POT pins")
# Create Label to read analog input
description_Label = Tkinter.Label(TOP, text="POT_input:- ")
description_Label.grid(column=1, row=1)
# Create Label to read analog input
analog_Read_Label = Tkinter.Label(TOP, text="Press_Start_process")
# Setting flag to toggle read option
FLAG = Tkinter.BooleanVar(TOP)
FLAG.set(True)
# Create Start button and associate with onStartButtonPress method
start_Button = Tkinter.Button(TOP, text="Start_reading",
   command=start_ button_press)
start_Button.grid(column=1, row=2)
# Create Stop button and associate with onStopButtonPress method
exit_Button = Tkinter.Button(TOP, text="Exit_reading",
   command=exit_button_press)
exit_Button.grid(column=2, row=2)
# Start and open the window
TOP.mainloop()
```

8.6.2 Tkinter GUI

Run the program described in Section 8.6.1 to create Tkinter GUI for reading a POT, and a window will appear (Figure 8.16). On pressing the "Start_reading" button, it will give readings of POT (Figure 8.17).

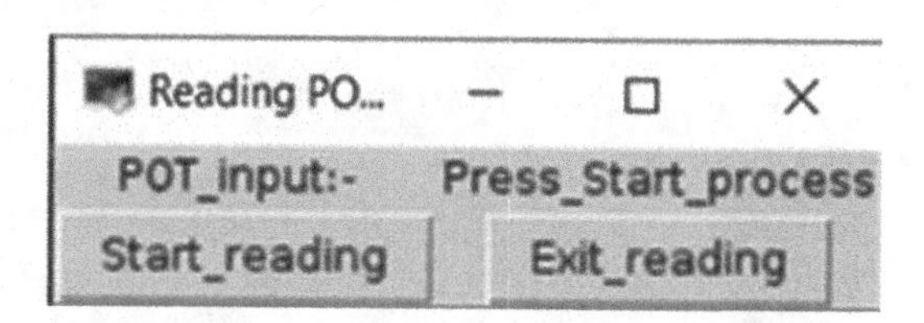

FIGURE 8.16
Tkinter GUI for reading a POT.

```
pi@raspberrypi:~ $ nano analog_in_tkinter.py
pi@raspberrypi:~ $ python analog_in_tkinter.py
analog values of pot
analog values of pot
analog values of pot
```

FIGURE 8.17
Window showing analog values of POT.

9

Data Acquisition with Python and Tkinter

9.1 Basics

The open() method: This function is a most commonly used method that is available in Python, and it is used to manipulate files.

To open a file command:

>>>f = open('test.txt', 'w')

This command creates a test.txt file in the same folder in which the Python interpreter is saved or the location from where the code is being executed.

The modes that can be used with the open() function are discussed in Table 9.1.

The write() method: Once the file is open in the writing mode, the write() method is used to start writing to the file object. The write() method takes input argument only in the format of a string.

>>>F.write("Hello World!\n")

Here "Hello World" is a string, and \n means a new line character.

To write a sequence of strings, the writelines() method is used:

>>> sq = ["Python programming for Arduino\n", "Bye\n"]

>>>W.writelines(sq)

The close() method: The close() method is used to close the file, and the file object can't be used again. The command is:

>>>W.close()

The read() method: This read() method reads the data of a file. To use this method, open the file with any of the reading compatible modes like: w+, r, r+, or a+:

>>>D = open('test.txt', 'r')

>>> D.read()

TABLE 9.1

Description of Modes

Mode	Description
W	This mode opens a file only to write. It overwrites an existing file.
w+	This mode opens a file to write and read both. It overwrites an existing file.
R	This mode opens a file only to read.
r+	This mode opens a file to write and read both.
A	This mode opens a file for appending, starting from end of the document.
a+	This mode opens a file for appending and reading, starting from end of the document.

```
'Hello World!\nPython programming for Arduino\nBye\n'
>>>D.close()
```

With this method, the entire contents of the file is stored into memory. To read the content line to line, use the readlines() method:

```
>>>D= open('test.txt', 'r')
>>>X =D.readlines()
>>> print X
['Hello World!\n', 'Python programming for Arduino\n', 'Bye\n']
>>>D.close()
```

9.2 CSV File

CSV files are used to store the data in Python. A CSV writer is used to write data on a CSV file and the reader to read it with simple commands:

CSV writer

```
import csv
data = [[1, 2, 3], ['a', 'b', 'c'], ['Python', 'Arduino', 'Programming']]
with open('example.csv', 'w') as f:
    w = csv.writer(f)
    for row in data:
        w.writerow(row)
```

CSV reader

```
import csv
with open('example.csv', 'r') as file:
    r = csv.reader(file)
    for row in r:
        print row
```

9.3 Storing Arduino Data with CSV File

The CSV files can be used to store the sensory data from Arduino. To understand the concept, a system is discussed. The system is comprised of Raspberry Pi, Arduino, a pyroelectric infrared (PIR) sensor (digital sensor), a potentiometer (POT) (as an analog sensor), and a power supply. The objective is to store the sensory data with CSV files and Arduino interfacing with Python. Figure 9.1 shows the circuit diagram of the system.

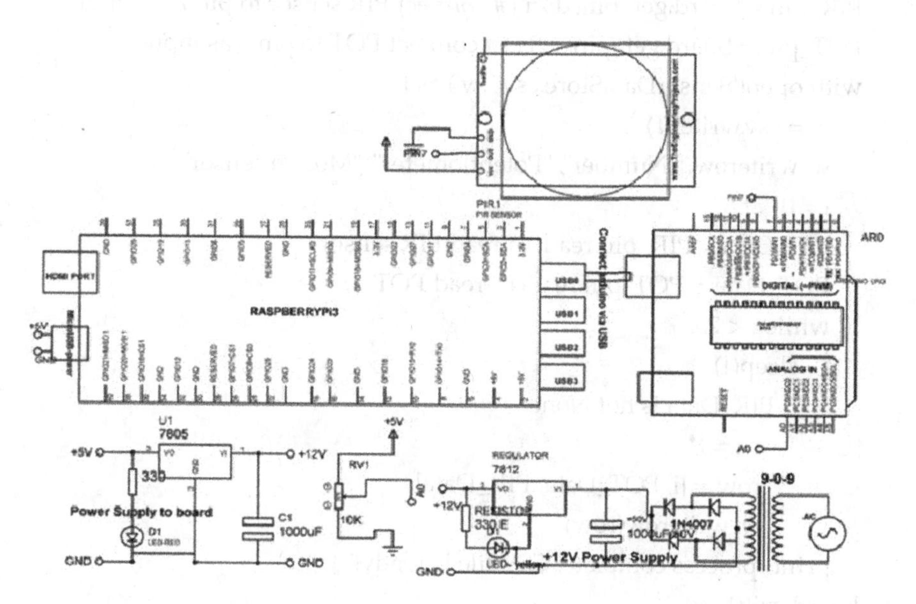

FIGURE 9.1
Circuit diagram to interface the PIR sensor and the POT with Arduino UNO and Pi.

Connections:

- Connect one terminal of POT to +5 V, the other terminal to ground, and wiper to pin (A0) of Arduino Uno.
- Connect pins (Vcc and ground) of the PIR sensor to +5 VDC and ground, respectively.
- Connect pin (OUT) of the PIR sensor to pin(7) of Arduino Uno.
- Connect Arduino Uni to Raspberry Pi through a USB.

9.3.1 Recipe

```
import csv # import CSV library
    import pyfirmata # import pyfirmata library
    import time as wait # import time library
    board = pyfirmata.Arduino('/dev/ttyUSB0')
    it = pyfirmata.util.Iterator(board)
    it.start() # start iterator
    PIR_pin = board.get_pin('d:7:i') # connect PIR sensor to pin 7 as input
    POT_pin = board.get_pin('a:0:i') # connect POT to pin 0 as input
    with open('SensorDataStore.csv', 'w') as f:
        w = csv.writer(f)
        w. writerow(["Number", "Potentiometer", "Motion sensor"])
        i = 0
        PIR_Data = PIR_pin.read() # read PIR sensor
        POT_Data = POT_pin.read() # read POT
        while i < 25:
            sleep(1)
            if PIR_Data is not None:
                i += 1
                row = [i, POT_Data, PIR_Data]
                w. writerow(row)
        print "process complete. CSV file is ready!"
    board.exit()
```

9.4 Plotting Random Numbers Using Matplotlib

The installation of matplotlib is an easy process on Ubuntu by a simple command:

$ sudo apt-get install python-matplotlib

Click on "yes" when prompted to install dependencies. The matplotlib library provides the plot() method to create line charts. The plot() method takes a list or an array data structure made up of integer or floating point numbers as input. Plot() uses values for the x-axis and y-axis, if two arrays are given as inputs. If only one list or array is provided as input, plot() assumes the values for the y-axis and autogenerates the incremental values for the x-axis:

pyplot.plot(x, y)

To change the style of line and makers with different colors, the plot() method can be used, e.g., for the solid line style command:

pyplot.plot(x, y, '-')

Figure 9.2 shows the plotting of the random number.

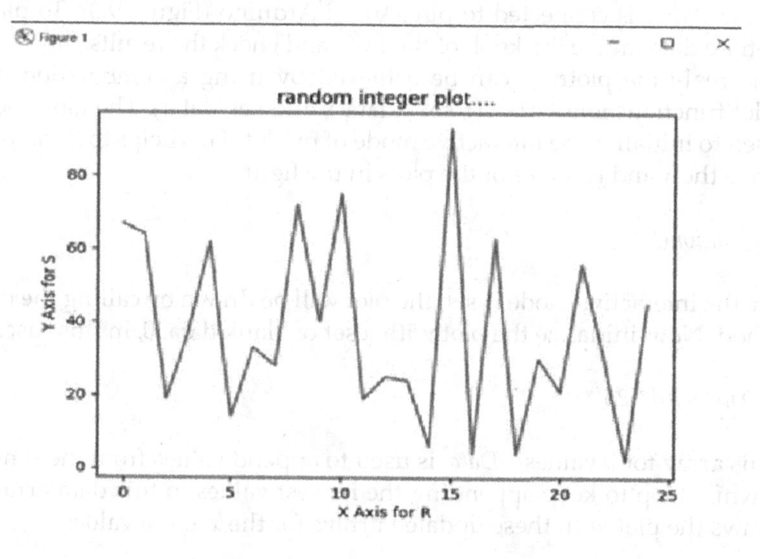

FIGURE 9.2
Plotting the random number.

9.4.1 Recipe

For generating and plotting random numbers:

```
import random
R = range(0,25)
S = [random.randint(0,100) for r in range(0,25)]
FIGURE1 = pyplot.figure()
pyplot.plot(R, S, '-')
pyplot.title('random integer plot....')
pyplot.xlabel('X Axis for R')
pyplot.ylabel('Y Axis for S')
pyplot.show()
```

9.5 Plotting Real-Time from Arduino

The real-time data plotting from Arduino is an important task where sensory data is critical. To understand the concept, real-time POT values are discussed in this section. The system is comprised of Raspberry Pi, Arduino, a POT, and a power supply. Arduino is connected to Raspberry Pi through a USB. One terminal of the POT is connected to +5 V, the other to the ground, and the wiper is connected to pin (A0) of Arduino (Figure 9.3). To plot the real-time data, move the knob of the POT and check the results.

The real-time plotting can be achieved by using a combination of the pyplot functions ion(), draw(), set_xdata(), and set_data(). The ion() method is used to initialize the interactive mode of pyplot. This helps to dynamically change the *x* and *y* values of the plots in the figure:

```
pyplot.ion()
```

Once the interactive mode is set, the plot will be drawn by calling the draw() method. Now, initialize the plot with a set of blank data, 0, in this case:

```
pData = [0] * 25
```

In this array for *y* values, *pData,* is used to append values from the sensor in the while loop to keep appending the newest values to this data array and redraws the plot with these updated arrays for the *x* and *y* values.

```
pData.append(float(a0.read()))
del pData[0]
```

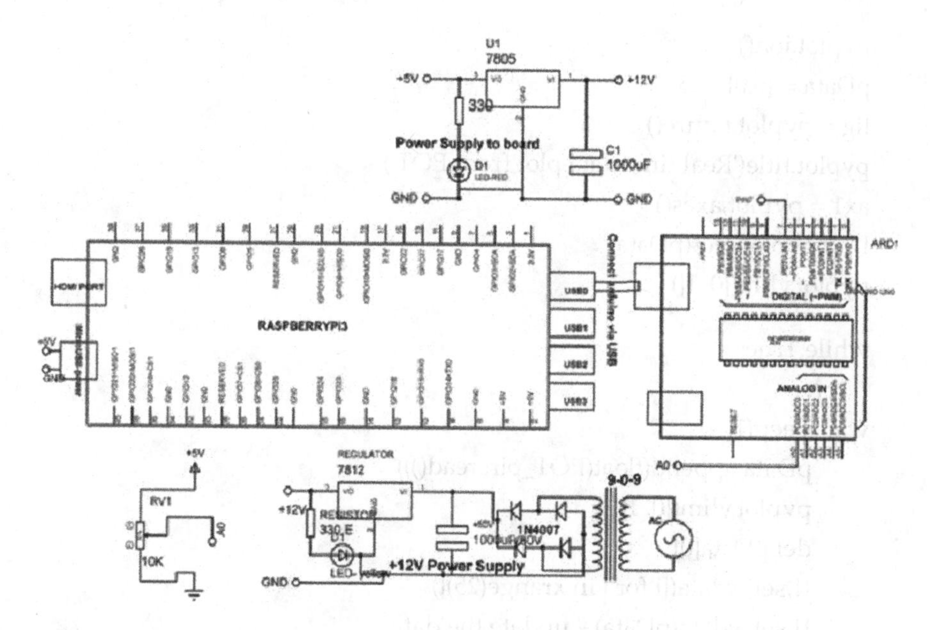

FIGURE 9.3
Circuit diagram for the interfacing of POT with Arduino.

The set_xdata() and set_ydata() methods are used to update the *x* and *y* axes data.

```
l1.set_xdata([i for i in xrange(25)])
l1.set_ydata(pData) # update the data
pyplot.draw() # update the plot
```

The code snippet, [i for i in xrange(25)], is to generate a list of 25 integer numbers that will start incrementally at 0 and end at 24.

9.5.1 Recipe

```
//from matplotlib import pyplot
    import pyfirmata # import pyfirmata library
    import time as wait
    board = pyfirmata.Arduino('/dev/ttyUSB0')
    wait.sleep(5) # wait for 5 Sec
    it = pyfirmata.util.Iterator(board)
    it.start() # start iterator
    POT_pin = board.get_pin('a:0:i') # connect POT to A0 pin as input
```

```
pyplot.ion()
pData = [0.0] * 25
fig = pyplot.figure()
pyplot.title('Real time data plot from POT')
ax1 = pyplot.axes()
l1, = pyplot.plot(pData)
pyplot.ylim([0, 1])

while True:
  try:
wait.sleep(1)
    pData.append(float(POT_pin.read()))
    pyplot.ylim([0, 1])
    del pData[0]
    l1.set_xdata([i for i in xrange(25)])
    l1.set_ydata(pData) # update the data
    pyplot.draw() # update the plot
  except KeyboardInterrupt:
    board.exit()
    break
```

9.6 Integrating the Plots in the Tkinter Window

Section 9.5 describes how to draw a plot for continuous sensory data from Arduino with the help of a POT. Python has a powerful integration capability with the matplotlib library and Tkinter graphical interface. For the same circuit as Figure 9.3, this integration is discussed. The program uses the interfacing of Tkinter with matplotlib.

9.6.1 Recipe

```
import sys # import sys library
from matplotlib import pyplot # import library
import pyfirmata # import library
import time as wait # import time library
import Tkinter # import library
```

```
    def start_button_press():
      while True:
          if FLAG.get():
  wait.sleep(1) # wait for 1 Sec
            pData.append(float(POT_pin.read()))
            pyplot.ylim([0, 1])
            del pData[0]
            l1.set_xdata([i for i in xrange(25)])
            l1.set_ydata(pData) # update the data
            pyplot.draw() # update the plot
            TOP.update()
          else:
            FLAG.set(True)
            break
def pause_button_press():
  FLAG.set(False)
def exit_button_press():
  print "out from data recording...."
pause_button_press()
  board.exit()
  pyplot.close(fig)
  TOP.quit()
  TOP.destroy()
  print "completed....."
  sys.exit()
board = pyfirmata.Arduino('/dev/ttyUSB0')
# Using iterator thread to avoid buffer overflow
it = pyfirmata.util.Iterator(board)
it.start() # start iterator
# Assign a role and variable to analog pin 0
POT_pin = board.get_pin('a:0:i') # read POT
# Tkinter canvas
TOP = Tkinter.Tk()
TOP.title("Tkinter + matplotlib")
# Create flag to work with indefinite while loop
FLAG = Tkinter.BooleanVar(TOP)
```

```
FLAG.set(True)
pyplot.ion()
pData = [0.0] * 25
figure = pyplot.figure()
pyplot.title('Potentiometer')
ax1 = pyplot.axes()
l1, = pyplot.plot(pData)
pyplot.ylim([0, 1])
# Create Start button and associate with start button press method
start_Button = Tkinter.Button(TOP, text="Start", command=start_
   button_press)
start_Button.grid(column=1, row=2)
# Create Stop button and associate with pause button press method
pause_Button = Tkinter.Button(TOP, text="Pause", command=pause_
   button_press)
# Create Exit button to exit from window
exit_Button=Tkinter.Button(TOP,text="Exit",command=exit_button_press)
exit_Button.grid(column=3, row=2)
TOP.mainloop()
```

Execute the program, and a window will appear on the screen (Figure 9.4). The plot can be controlled by using "Start", "Pause" and "Exit" buttons. Click on start button and rotate the knob of POT and see the changes in plot. The process can be paused or close the program with "Exit" button.

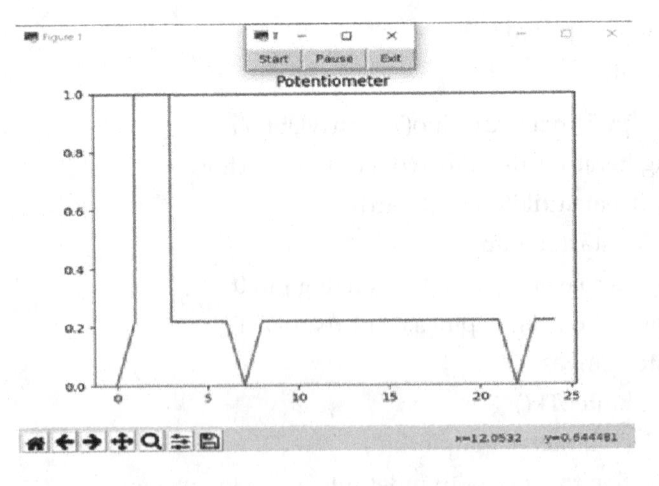

FIGURE 9.4
Plot for real-time data from Arduino.

Section IV

Connecting to the Cloud

10

Smart IoT Systems

10.1 DHT11 Data Logger with ThingSpeak Server

The objective of this project is to capture the real-time data of DHT11 by Raspberry Pi and upload to the cloud. DHT11 is a temperature and humidity sensor. ThingSpeak is used as a cloud server for data forecasting. The process involves three parts: setting up Raspberry Pi, configuring Adafruit library, and then connecting to ThingSpeak. The system is comprised of a Raspberry Pi, a power supply, an Arduino Uno, and a DHT11 sensor. Connect pin (Vcc) of DHT11 to +5V and ground, respectively. Connect pin(OUT) of DHT11 to GPIO17 of the Raspberry Pi (Figure 10.1).

10.1.1 Installation of DHT11 Library

Install the DHT11 library with following commands:

#git clone https://github.com/adafruit/Adafruit_Python_DHT.git.... // Enter the command to clone the library

#cd Adafruit_Python_DHT // Installed directory using this command

#sudo apt-get install build-essential python-dev //...Now download the required modules using the command

#sudo python setup.py install //....Then install the library using the command.

10.1.2 Steps to Create a Channel in ThingSpeak

1. Sign in to ThingSpeak by creating a new MathWorks account.
2. Click Channels >MyChannels (Figure 10.2).
3. Click on "New Channel" and a window will open (Figure 10.3).

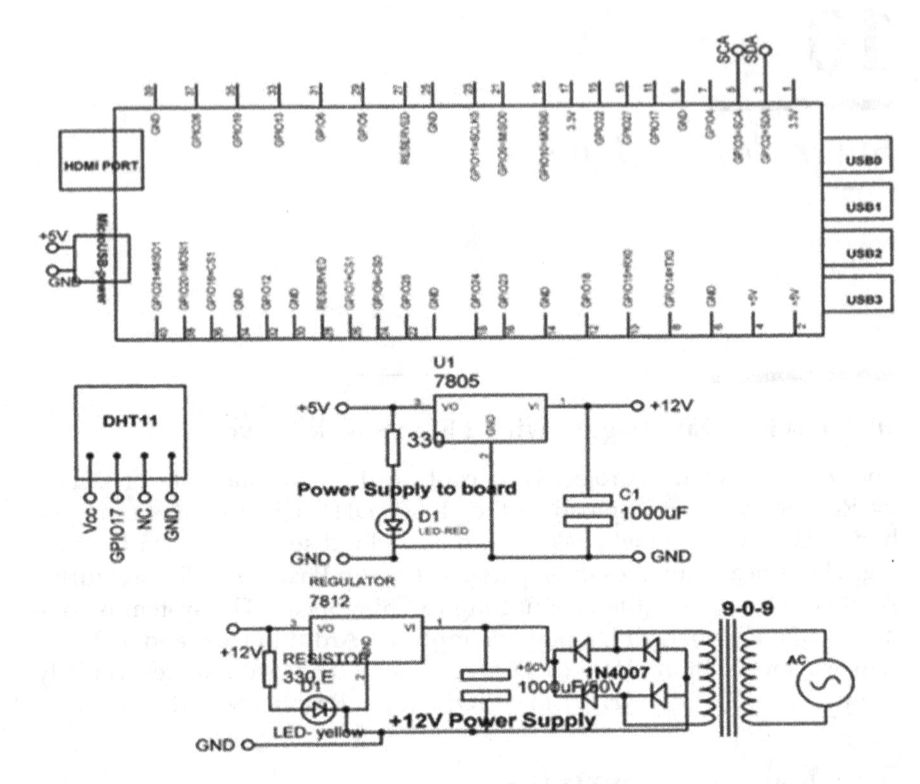

FIGURE 10.1
Circuit diagram for interfacing of DHT11 with Raspberry Pi.

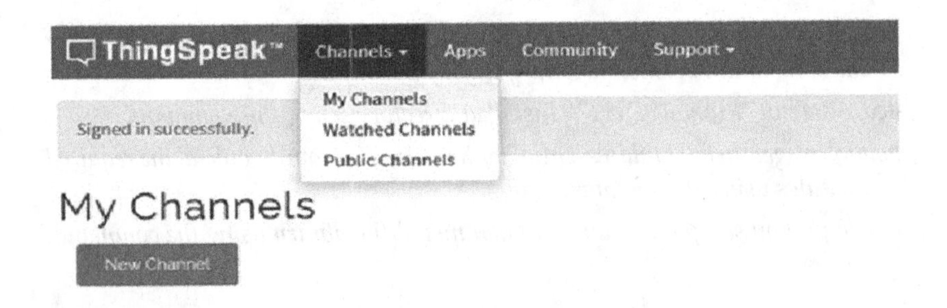

FIGURE 10.2
Window for ThingSpeak.

4. Enter the channel setting values and click on "Save Channel" at the bottom of the settings.
5. Check API write key (this key needs to write in the program).
6. Write the program described in Section 10.1.3 and check the sensory data on cloud server (Figures 10.4 and 10.5).

Name	Created	Updated At
Channel 293693	2017-06-26	2017-09-20 04:23
Private · Public · Settings · Sharing · API Keys · Data Import / Export		

FIGURE 10.3
New channel in my channels.

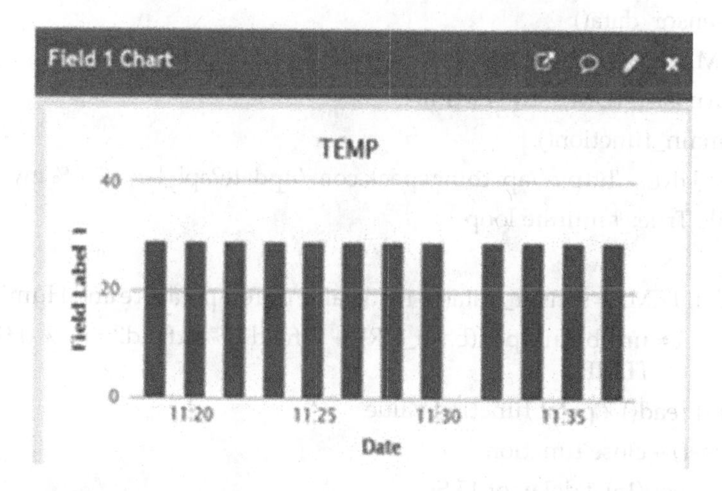

FIGURE 10.4
Graph for temperature data.

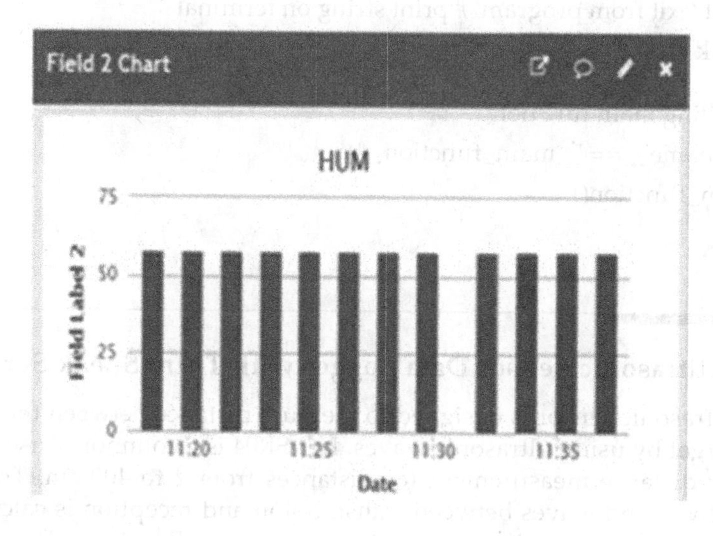

FIGURE 10.5
Graph for humidity data.

10.1.3 Recipe

```
import sys # import sys library
importRPi.GPIO as RAJ # import GPIO library
import time as wait # add time library
importAdafruit_DHT as RAJ_DHT
import urllib2 #import urllib library
my_API = 'F1MAEF943TLMVTC1'
defsensor_data():
HUM, TEMP= RAJ_DHT.read_retry(RAJ_DHT.DHT11, 21)
return (str(HUM), str(TEMP))
defmain_function():
base_URL = 'https://api.thingspeak.com/update?api_key=%s' % my_API
while True: # infinite loop
try:
HUM, TEMP = sensor_data() # read value of Temperature and Humidity
        f = urllib2.urlopen(base_URL + '&field1=%s&field2=%s' % (HUM,
          TEMP))
printf.read() # print function value
f.close() # close function
wait.sleep(15) # delay of 15 Sec
except:
print 'exit from program' # print string on terminal
break

# calling main function
if __name__ == '__main_function__':
main_function()
```

10.2 Ultrasonic Sensor Data Logger with ThingSpeak Server

The ultrasonic sensor is designed to measure distance between the source and target by using ultrasonic waves. HC-SR04 is a commonly used module for distance measurement for distances from 2 to 400 cm. The time taken by sound waves between transmission and reception is calculated. To measure distance, the system is comprised of a Raspberry Pi, a power supply, and an ultrasonic sensor. Connect pin (Vcc) and pin (GND) of the

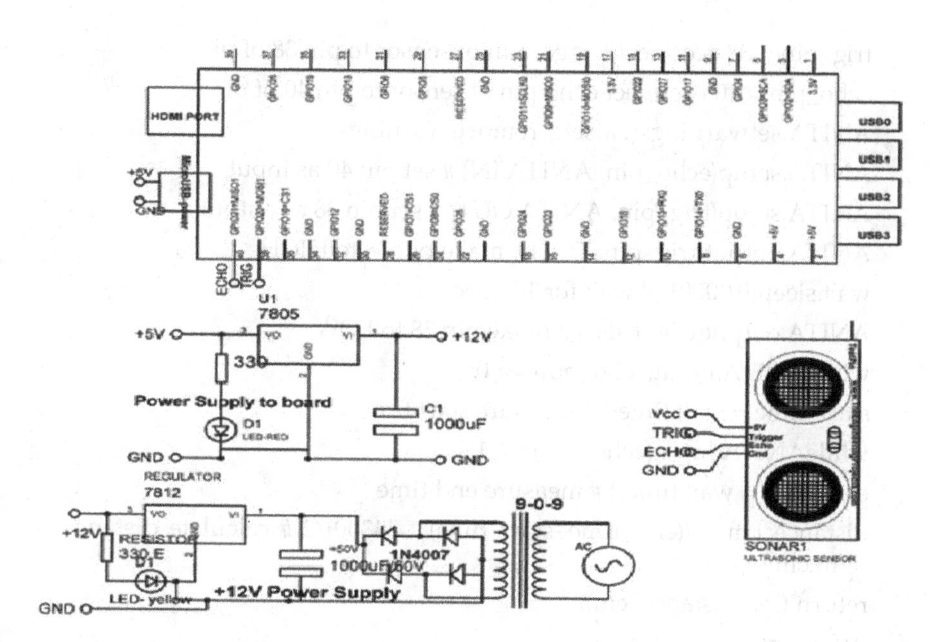

FIGURE 10.6
Circuit diagram for interfacing of ultrasonic sensor with Raspberry Pi.

ultrasonic sensor to +5 V and ground, respectively. Connect pin(TRIG) and pin(ECHO) of the ultrasonic sensor to GPIO20 and GPIO(21) of Raspberry Pi, respectively (Figure 10.6). Provide a "HIGH" signal to "TRIG" input, at least for 10-μS duration. This process will enable the module to transmit eight 40-KHz ultrasonic waves. If any obstacle is in the way of the waves, these are reflected back to the "ECHO" pin. The time taken by pulse is actually for to-and-fro travel of ultrasonic signals, so time is taken as Time/2. Distance = Speed * Time/2; here speed of sound at sea level = 343 m/s or 34,300 cm/s. Follow the steps described in Section 10.1.2 to create the ThingSpeak server, write the program described in Section 10.2.1, and check the results.

10.2.1 Recipe

```
import time as wait # add time library
importRPi.GPIO as ANITA # add pi GPIO library
import sys # add sys library
import urllib2 # add urllib
my_API = 'F1MAEF943TLMVTC1'
defget_sensor_data():
ANITA.setmode(ANITA.BOARD) # use pi in board mode
```

```
trig_pin = 38 #connect trigger pin of sensor to pin 38 of pi
echo_pin = 40 #connect echo pin of sensor to pin 40 of pi
ANITA.setwarnings(False) # remove warnings
ANITA.setup(echo_pin, ANITA.IN) # set pin 40 as input
ANITA.setup(trig_pin, ANITA.OUT) # set pin 38 as output
ANITA.output(trig_pin, True) # make pin 38 to HIGH
wait.sleep(0.00001) # wait for 10 uSec
ANITA.output(trig, False) # make pin 38 to LOW
whileANITA.input(echo_pin) == 0:
start_time = wait.time() # measure start time
whileANITA.input(echo_pin) == 1:
end_time = wait.time() # measure end time
distance_cm = ((end_time–start_time) * 34300)/2 # calculate distance
  in cm
return (int (distance_cm))
def main():
base_URL = 'https://api.thingspeak.com/update?api_key=%s' % my_API
while True:
try:
distance_cm = get_sensor_data() # read sensor
        f=urllib2.urlopen(base_URL+"&field3=%s"%int(distance_cm/5))
printf.read() # print function value
f.close() # close function
wait.sleep(15) # wait for 15 Sec
except:
print 'exit from main program' # print string on terminal
break
```

10.3 Air Quality Monitoring System and Data Logger with ThingSpeak Server

Air pollution is the largest environmental issue in the world today. Air pollution leads to many adverse effects on human health, climate, and the ecosystem. This generates a need for measurement and analysis of real-time air quality. Air Quality Index (AQI) is an important parameter for air quality; it is a standard (Table 10.1).

TABLE 10.1
AQI Classification

AQI Range	AQI Category
0–50	Excellent
51–100	Good
100–150	Lightly polluted
151–200	Moderately polluted
201–250	Heavily polluted
251–300	Severely polluted

To understand air quality monitoring, a system is designed with air quality sensors and a camera. Here, a camera is connected to capture the images of a polluted area. Figure 10.7 shows the block diagram of the system. It is comprised of a Raspberry Pi, a MQ2 (smoke detector), a MQ7 (CO gas detector), a DHT11 (temperature/humidity sensor), a camera, a power supply, a dust sampler (GP2Y1030), and an Arduino Uno.

Connect the components as shown in Figure 10.8. Follow the steps described in Section 10.1.2 to create the ThingSpeak server, write the program described in Section 10.3.1, and check the results as shown in Figures 10.9 through 10.12.

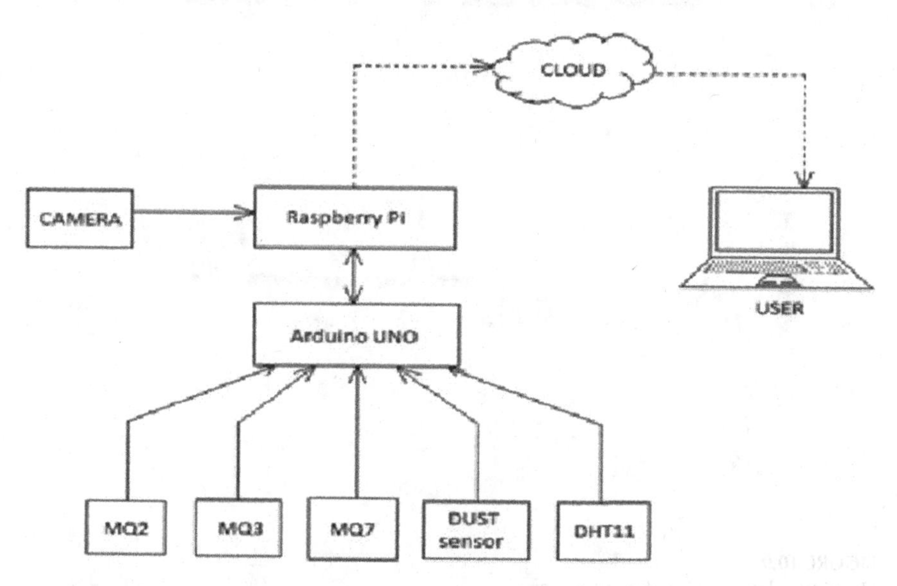

FIGURE 10.7
Block diagram of the air quality monitoring system.

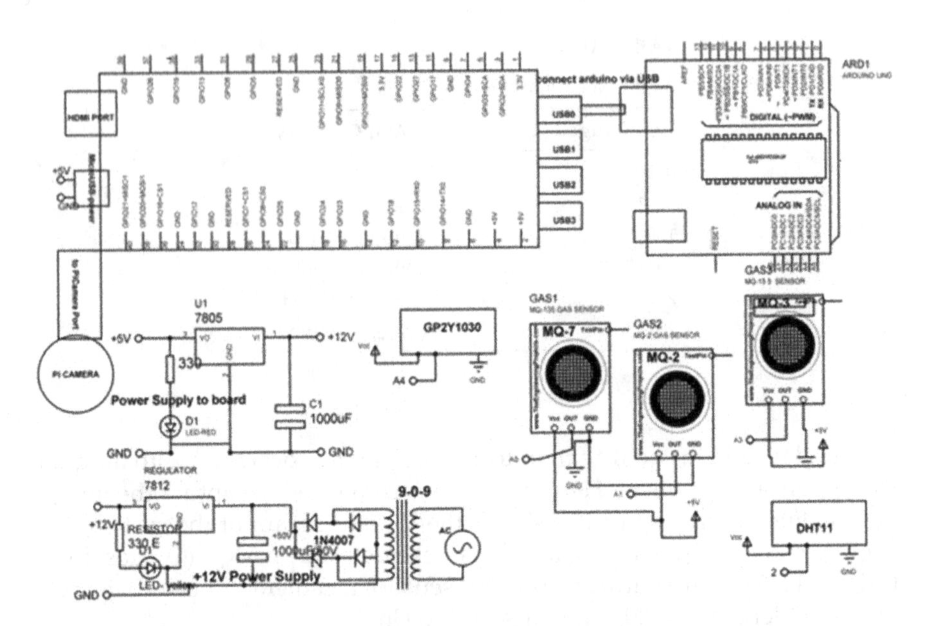

FIGURE 10.8
Circuit diagram for air quality monitoring system.

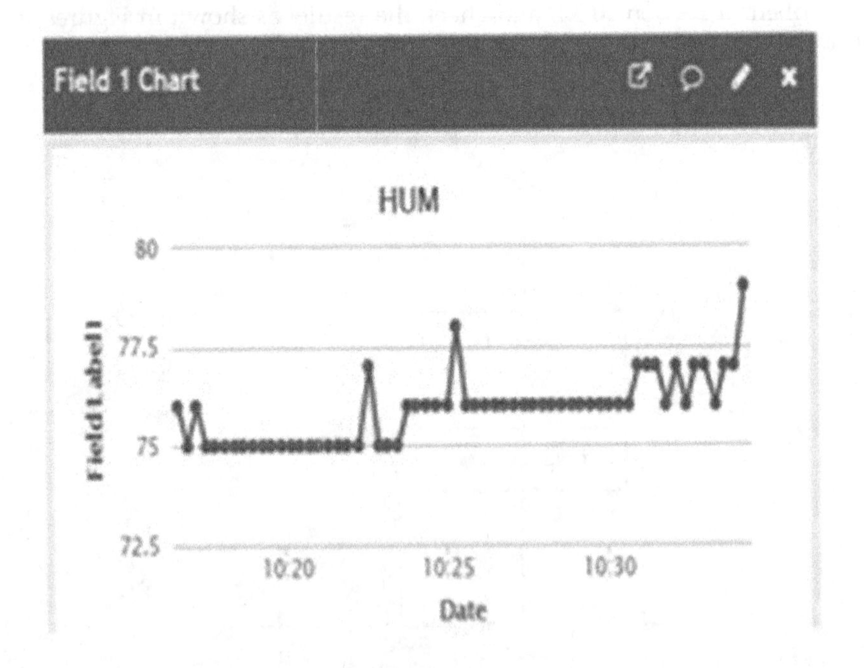

FIGURE 10.9
Humidity data on ThingSpeak server.

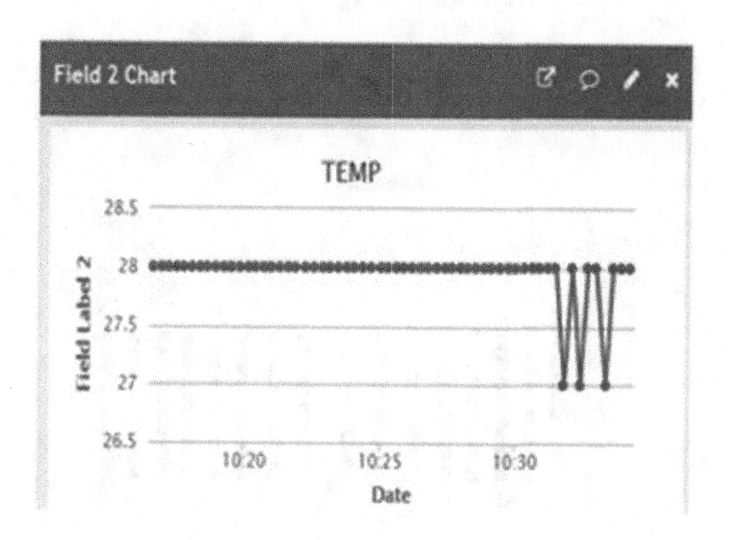

FIGURE 10.10
Temperature data on ThingSpeak server.

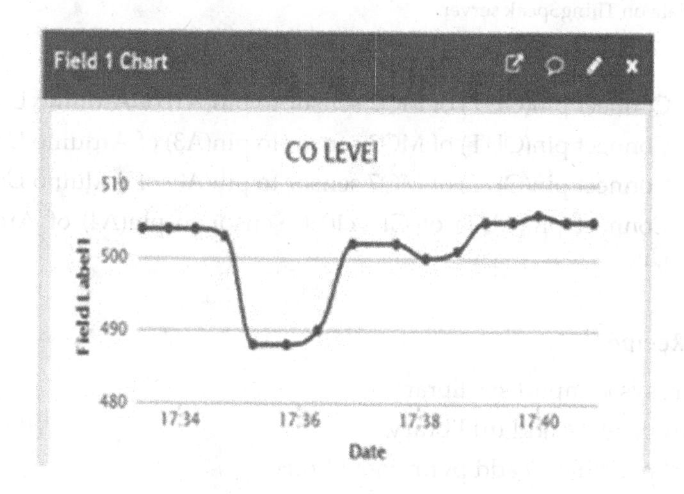

FIGURE 10.11
CO level data on ThingSpeak server.

Connections

- Connect camera to the dedicated port at Raspberry Pi.
- Connect pins(Vcc and ground) of MQ2, MQ3, MQ7, GP2Y1030, and DHT11 sensor to +5 VDC and ground, respectively.
- Connect pin(OUT) of DHT11 sensor to pin(2) of Arduino Uno.
- Connect Arduino Uni to Raspberry Pi through a USB.

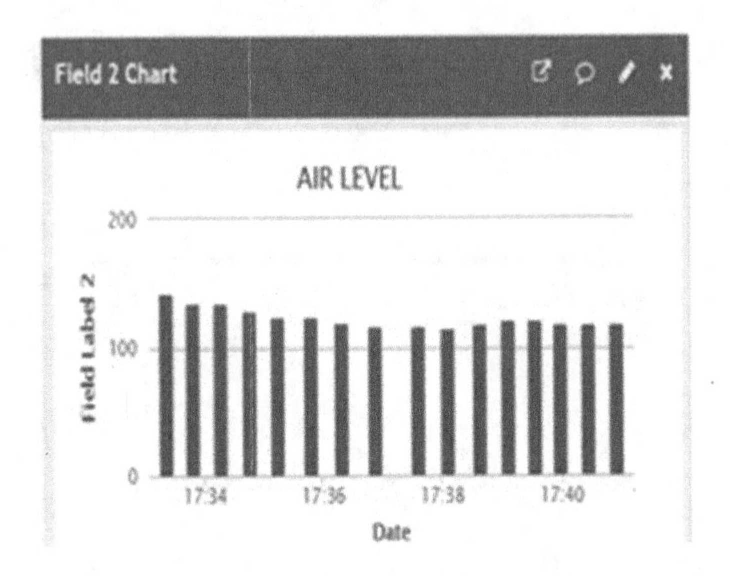

FIGURE 10.12
Air level data on ThingSpeak server.

- Connect pin(OUT) of MQ2 sensor to pin(A1) of Arduino Uno.
- Connect pin(OUT) of MQ3 sensor to pin(A3) of Arduino Uno.
- Connect pin(OUT) of MQ7 sensor to pin(A0) of Arduino Uno.
- Connect pin(OUT) of GP2Y1030 sensor to pin(A4) of Arduino Uno.

10.3.1 Recipe

```
import sys # import sys library
import urllib2 # add url library
import pyfirmata # add pyfirmata library
importAdafruit_DHT as RAJ_DHT
import time as wait # import time library
importRPi.GPIO as RAJ # import ri GPIO library
frompicamera import PiCamera # import picamera library
board = pyfirmata.Arduino('/dev/ttyUSB0') # USB com number
led_Power = board.get_pin('d:12:o') # connect led_power pin to 12 pin
    of Arduino as output
```

```
measure_Pin = board.get_pin('a:5:i') # connect senor pin to A5 pin of
   Arduino as input
co_Pin = board.get_pin('a:0:i') # connect co sensor pin to A0 pin of
   Arduino as input
it = pyfirmata.util.Iterator(board) # use iterator
it.start() # start iterator
measure_Pin.enable_reporting() # enable pin
co_Pin.enable_reporting() # enable pin
RAJ_camera = PiCamera()
photo_Count = 0 # assume initial value
timer = 5 # assume initial value
previous_Dust_Value = 0 # assume initial value
sampling_Time = 0.000280 # assume initial value
delta_Time = 0.000040 # assume initial value
sleep_Time = 0.009680 # assume initial value

co_Measured = 0 # assume initial value
vo_Measured = 0 # assume initial value
calc_Voltage = 0 # assume initial value
dust_Density = 0 # assume initial value
HUM = 0 # assume initial value
TEMP = 0 # assume initial value
my_API = 'SRWU0GUE90JLEOP2'
base_URL = 'https://api.thingspeak.com/update?api_key=%s' % my_API
defget_Temp_Hum(): # function to get temperature and humidity data
HUM, TEMP = RAJ_DHT.read_retry(RAJ_DHT.DHT11,17)
defget_Dust_Data(): # function to get dust data
led_Power.write(0) #make led_Power pin LOW
wait.sleep(sampling_Time) # wait
vo_Measured = measure_Pin.read()
wait.sleep(delta_Time) # wait
led_Power.write(1) # make led_Power pin HIGH
wait.sleep(sleep_Time) # wait
calc_Voltage = vo_Measured * (5.0)
dust_Density = (0.17) * calc_Voltage - (0.1)
```

```
if (dust_Density< 0):
dust_Density = 0.00

deftake_Picture(): # function to take picture
if (dust_Density != previous_Dust_Value):
RAJ_camera.start_preview()
wait.sleep(2) # wait for 2 sec
RAJ_camera.capture('/home/pi/Desktop/image%s.jpg' % photo_Count)
RAJ_camera.stop_preview() # stop camera preview
Photo_Count += 1
Previous_Dust_Value = dust_Density
timer = 5
else:
if( timer == 0):
RAJ_camera.capture('/home/pi/Desktop/image%s.jpg' % photo_Count)
Photo_Count += 1
timer = 5
else:
timer -= 1
defget_Co(): # function to measure CO
co_Measured = co_Pin.read() # read pin
co_Measured = co_Measured * 1000 # scaling
defget_String_Format(): # format data in string
   return(str(TEMP) ,str(HUM), str(dust_Density), str(co_Measured)))
while True: # infinite loop
      get_Temp_Hum() # call function
      get_Co() # call function
      get_Dust_Data() # call function
      if (dust_Density>= 0.5):
      take_Picture()
         T,RH,D,CO = get_String_Format()
         f=urllib2.urlopen(baseURL+'&field1=%s&field2=%s&field4=%
            s&field4=%s' % (T,RH,D,CO))
      printf.read() # print function
      f.close() # close function
      wait.sleep(15) # 15 sec delay
```

10.4 Landslide Detection and Disaster Management System

Landslides are common in the rainy season in the hilly areas of India. By predicting landslides, many human lives can be saved. Natural disasters like landslides and flooding are caused by global warming, which may result in human and economic losses. A system for landslide detection and disaster management detects the landsliding and also the presence of human in the affected area. A buzzer is used to alert for the presence of human, is also a part of the system. Human presence can be detected with a PIR sensor and with a camera. The complete system is comprised of a Raspberry Pi, a Pi camera, a PIR (motion detector) sensor, a BMP180 (pressure sensor), a DHT11 (temperature and humidity sensor), a rain sensor, an accelerometer (tilt detector), a buzzer, an Arduino Uno, and a power supply. Connect the components as shown in Figures 10.13 through 10.17, which show the data on the ThingSpeak server.

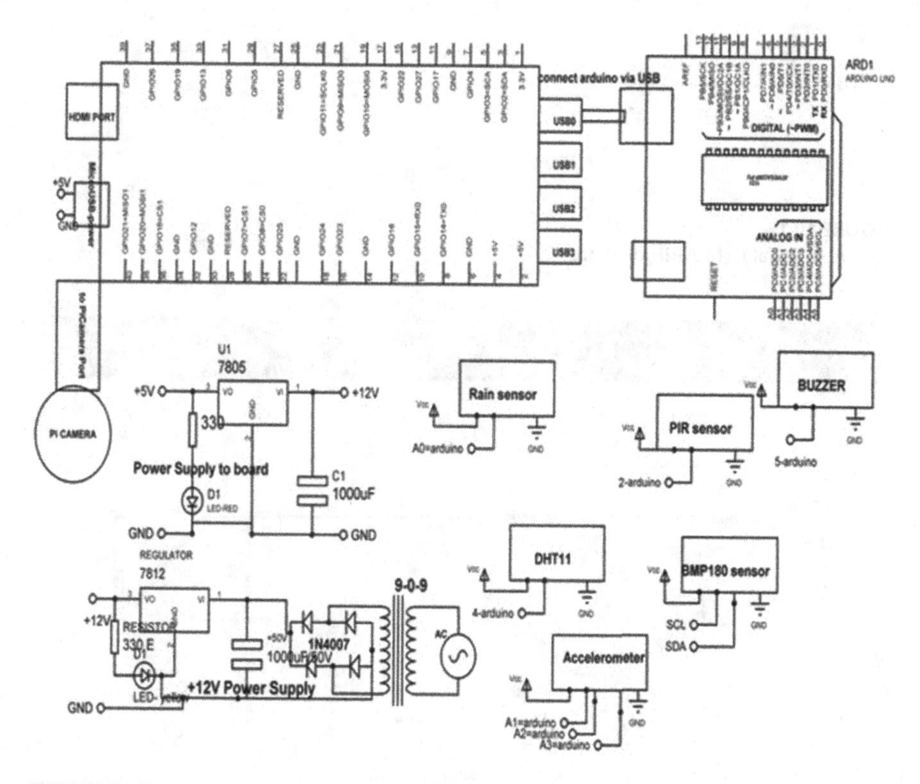

FIGURE 10.13
Circuit diagram for landslide detection and disaster management system.

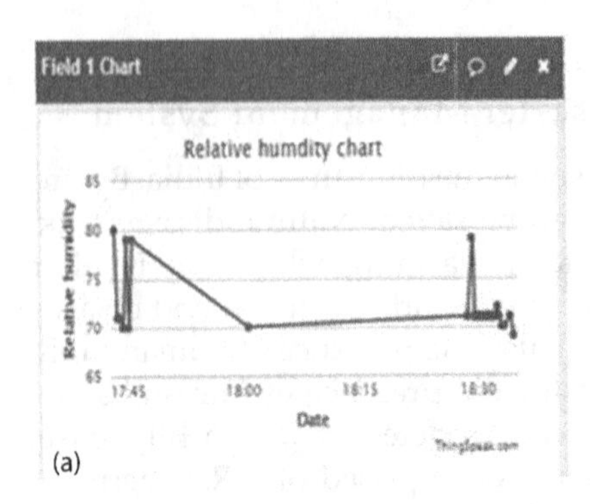

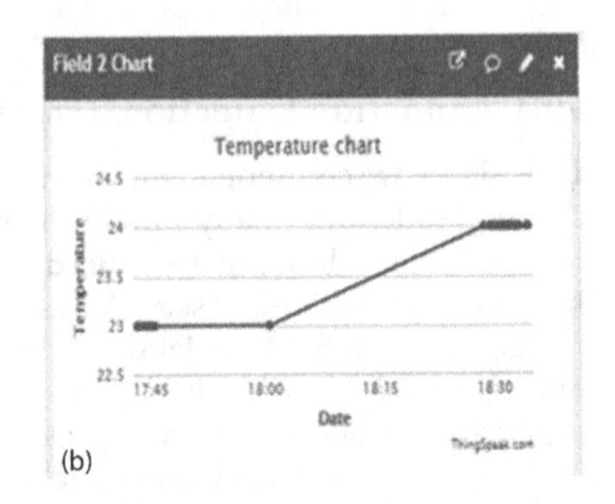

FIGURE 10.14
(a) Humidity data. (b) Temperature data.

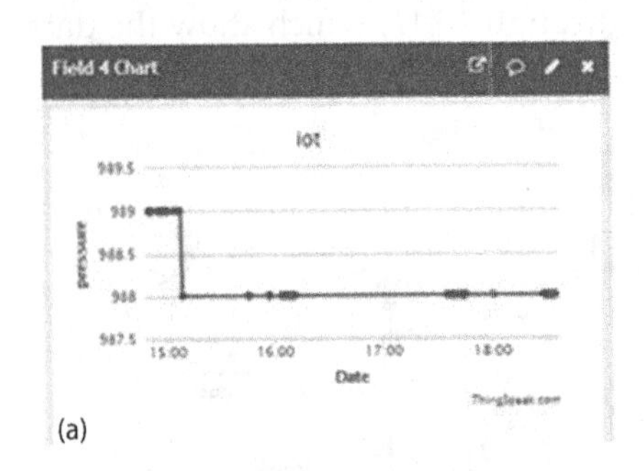

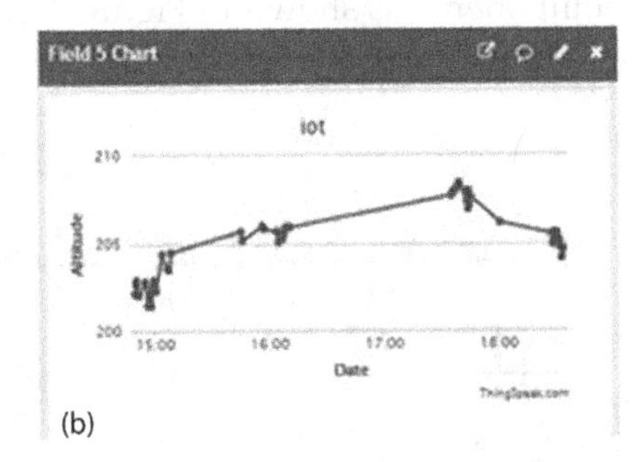

FIGURE 10.15
(a) Pressure data. (b) Altitude data.

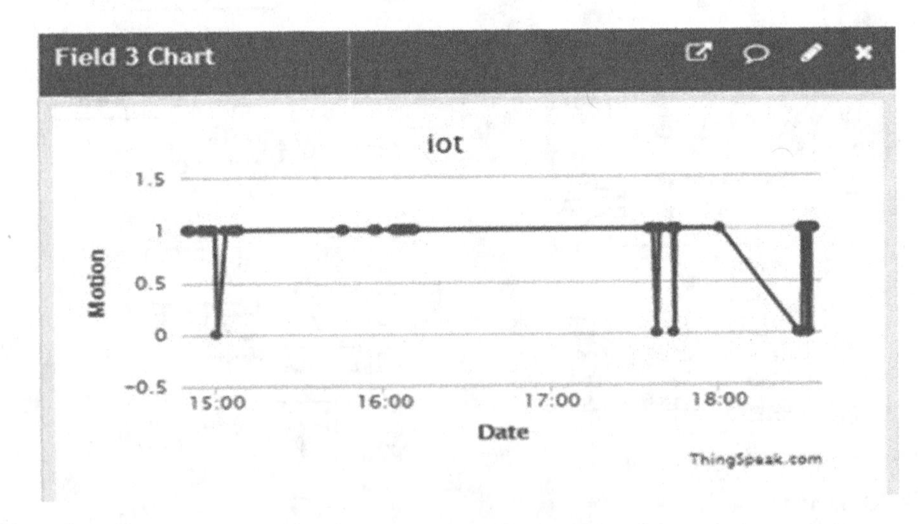

FIGURE 10.16
PIR sensor data.

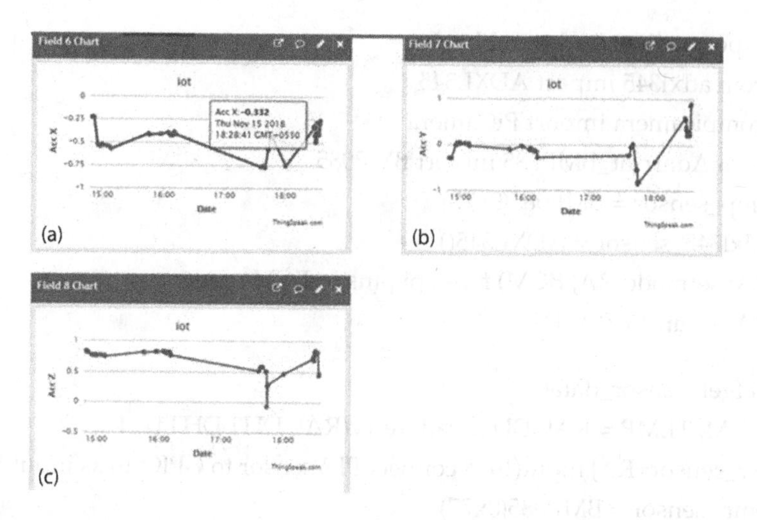

FIGURE 10.17
(a) Accelerometer *x*-axis data. (b) *y*-axis data. (c) *z*-axis data.

Connections

- Connect camera to the dedicated port at Raspberry Pi.
- Connect pins (Vcc and ground) of the PIR sensor, BMP180 sensor, rain sesnor, accelerometer, buzzer, and DHT11 sensor to +5 VDC and ground, respectively.
- Connect pin(OUT) of DHT11 sensor to pin(4) of Arduino Uno.
- Connect Arduino Uni to Raspberry Pi through a USB.
- Connect pin(OUT) of the PIR sensor to pin(2) of Arduino Uno.
- Connect pin(OUT) of the rain sensor to pin(A0) of Arduino Uno.
- Connect pin(OUT) of the buzzer to pin(5) of Arduino Uno.
- Connect pin(SCL) of BMP180 to pin(A5) of Arduino Uno.
- Connect pin(SDA) of BMP180 to pin(A4) of Arduino Uno.
- Connect "x axis" of accelerometer to pin (A1) of Arduino Uno.
- Connect "y axis" of accelerometer to pin (A2) of Arduino Uno.
- Connect "z axis" of accelerometer to pin (A3) of Arduino Uno.

10.4.1 Recipe

```
import sys # import library
importRPi.GPIO as RAJ
import time as wait # import time library
importAdafruit_DHT as RAJ_DHT
```

```
import urllib2 # add url library
from adxl345 import ADXL345
frompicamera import PiCamera
from Adafruit_BMP085 import BMP085
bmp_sensor = BMP085(0x77)
adxl345_sensor = ADXL345()
RAJ.setmode(RAJ.BCM) # use pi pins in BCM mode
RAJ.setup(16, RAJ.IN)

defget_sensor_data():
HUM, TEMP = RAJ_DHT.read_retry(RAJ_DHT.DHT11, 4)
pir_sensor=RAJ.input(16) # connect PIR sensor to GPIO 16 as input
bmp_sensor = BMP085(0x77)
   pressure1= bmp.readPressure() # read pressure
pressure_data=pressure1/100 # scale pressure value
altitude_data = bmp.readAltitude() # read altitude
axes = adxl345.getAxes(True)
   x1_data= axes['x']
   y1_data=axes['y']
   z1_data=axes['z']
return (str(HUM), str(TEMP), str(pir_sensor), str(pressure_data),
   str(altitude_data), str(x1_data), str(y1_data), str(z1_data))

defmain_program():
count=0 # initialise count
   pir1_data=GPIO.input(16) # connect PIR sensor to GPIO16 as input
iflen(sys.argv) < 2:
print('print data') # print string on terminal
exit(0)
print 'process start...' # print string on terminal
base_URL = 'https://api.thingspeak.com/update?api_key=%s' % sys.
   argv[1]

while True:
try:
HUM, TEMP, pir_sensor, pressure_data, altitude_data, x1_data, y1_
   data, z1_data =
get_sensor_data()
```

```
        f = urllib2.urlopen(base_URL +"&field1=%s&field2=%s&field3=%
            s&field4=%s&field5=%s&field6=%s&field7=%s&field8=%s"
            % (HUM, TEMP, pir_sensor, pressure_data, altitude_data,
            x1_data, y1_data, z1_data))
printf.read() # read function

if (pirl_data==1):
print "hello" # print string on terminal
RAJ_camera = PiCamera()
RAJ_camera.start_preview()
wait.sleep(5) # delay of 5 sec
count =count+ 1
RAJ_camera.capture('img/image%s.jpg' % count) # capture image
RAJ_camera.stop_preview() # stop preview
RAJ_camera.close() # close camera
f.close() # close function
wait.sleep(15) # dely 15 sec
except:
print 'exit from main program' # print string on terminal
break
if __name__ == '__main_program__':
main_program()
```

10.5 Smart Motion Detector and Upload Image to gmail.com

A smart motion detector is a system where if a motion is detected, then a picture is captured and communicated to the predefined "Gmail" address of the authorized person. The system is comprised of a Raspberry Pi, a Pi camera, a PIR sensor, an LED (as indicator), and a power supply. Connect the components as shown in Figure 10.18.

Connections:

- Connect the camera to the dedicated port at Raspberry Pi.
- Connect pins(Vcc and ground) of the PIR sensor to +5 VDC and ground, respectively.
- Connect pin(OUT) of the PIR sensor to GPIO21 of Raspberry Pi.
- Connect the cathode terminal of the LED to the ground, and the anode terminal to GPIO13 of Raspberry Pi through a resistor.

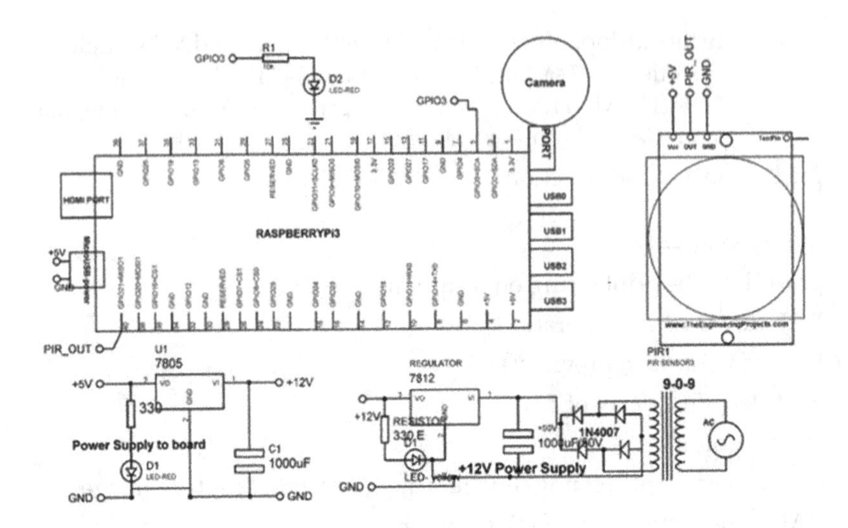

FIGURE 10.18
Circuit diagram for smart motion detector.

10.5.1 Configuring Raspberry Pi with Camera and Gmail

Step 1: Update and upgrade the Raspberry Pi with the latest packages by using following commands:

sudo apt-get update
sudo apt-get upgrade
or
sudo apt-get install python3 //for Python 3.0

Step 2: Install and configure an SMTP service.

sudo apt-get install ssmtp
sudo nano /etc/ssmtp/ssmtp.conf

Step 3: Enable the permissions for SSH and camera in Raspberry Pi configuration.

Step 4: Allow "Gmail" SMTP access with standard authentication by following the steps:

1. Login to your Gmail account with username and password.
2. Go to "My Account" at the top-right corner.
3. Select the "Sign-in & security" section locate "Connected apps & sites."
4. Turn "ON" the settings for "Allow less secure apps."

10.5.2 Recipe

```
importRPi.GPIO as ANITA # import pi GPIO
importpicamera as RAJ_camera # import camera library
import time as wait # import time library
importsmtplib # import smtp library
fromemail.mime.multipart import MIMEMultipart # add library
fromemail.mime.text import MIMEText # add library
fromemail.mime.base import MIMEBase # add library
from email import encoders # add library
fromemail.mime.image import MIMEImage # add library
from_addr = "sssss@gmail.com" # change the email address accordingly
to_addr = "rrrrrr@gmail.com"
mail_RAJ = MIMEMultipart()
mail_RAJ['From'] = from_addr
mail_RAJ['To'] = to_addr
mail_RAJ['Subject'] = "Attachment"
body = "Please find the attachment"
led_pin=3 # connect led to pin 3 of raspberry pi GPIO
pir_pin=21 # connect PIR sensor to pin 21 of raspberry pi GPIO

HIGH=1 # assign value
LOW=0 # assign value
RAJ.setwarnings(False) # remove warnings
RAJ.setmode(RAJ.BCM) # use pins in BCM mode
RAJ.setup(led_pin, RAJ.OUT) # set direction of pin to OUTPUT
RAJ.setup(pir_pin, RAJ.IN) # set direction of pin to INPUT
data=""

defsend_Mail(data):
mail_RAJ.attach(MIMEText(body, 'plain'))
print data # print data on terminal
dat='%s.jpg'%data
print (data) # print data on terminal
attachment_RAJ = open(dat, 'rb')
image_RAJ=MIMEImage(attachment_RAJ.read())
attachment_RAJ.close()
```

```
mail_RAJ.attach(image_RAJ) # attach image
server_RAJ = smtplib.SMTP('smtp.gmail.com', 587)
server_RAJ.starttls()
server_RAJ.login(from_addr, "ssss@gmail.com")
text = mail_RAJ.as_string()
server_RAJ.sendmail(from_addr, to_addr, text)
server_RAJ.quit() # quit from server

defRAJ_capture_image():
data= wait.strftime("%d_%b_%Y|%H:%M:%S")
RAJ_camera.start_preview() # start preview
wait.sleep(5) # wait for 5 Sec
print data # print on terminal
RAJ_camera.capture('%s.jpg'%data)
RAJ_camera.stop_preview()
wait.sleep(1) # wait for 1 Sec
send_Mail(data) # send mail

RAJ.output(led_pin, 0) # make led_pin to LOW
RAJ_camera = picamera.PiCamera()
RAJ_camera.rotation=180 # rotate camera
RAJ_camera.awb_mode= 'auto'
RAJ_camera.brightness=55 # choose brightness level
while 1: # infinite loop
ifRAJ.input(pir)==1: # check the state of PIR sensor
RAJ.output(led_pin, HIGH) # make led_pin to HIGH
RAJ_capture_image() # capture image
while(RAJ.input(pir_pin)==1): # check state of PIR sensor
wait.sleep(1) # delay of 1 sec
else:
gpio.output(led, LOW) # make led_pin to LOW
wait.sleep(1) # delay of 1 sec
```

11

Blynk Application with Raspberry Pi

11.1 Introduction to Blynk

There are three major components in the platform:

Blynk App—Allow to create an interface by using various widgets.

Blynk Server—It is responsible for the communication between the smartphone and hardware. It is an open-source platform.

Blynk Library—Blynk can work with Arduino, Raspberry Pi, or on similar boards.

Blynk works with any model of Raspberry Pi through Wi-Fi/Ethernet.

11.1.1 Installing Blynk on Raspberry Pi

To install Blynk on Raspberry Pi, first install the Blynk libraries. It is required to have the latest version of Raspbian to install Blynk on Pi.

```
sudo apt-get update
sudo apt-get install git-core // If GIT is not installed, install it
git clone git://git.drogon.net/wiringPi // Install WiringPi using GIT
cdwiringPi // go to wiringPi and run the build.
. /build
```

To install Blynk globally, run the commands:

```
git clone https://github.com/blynkkk/blynk-library.git
cd blynk-library/linux
make clean all target=raspberry
Sudo./blynk --token=token num // authentication token
```

Figure 11.1 shows the window after the execution of the commands.

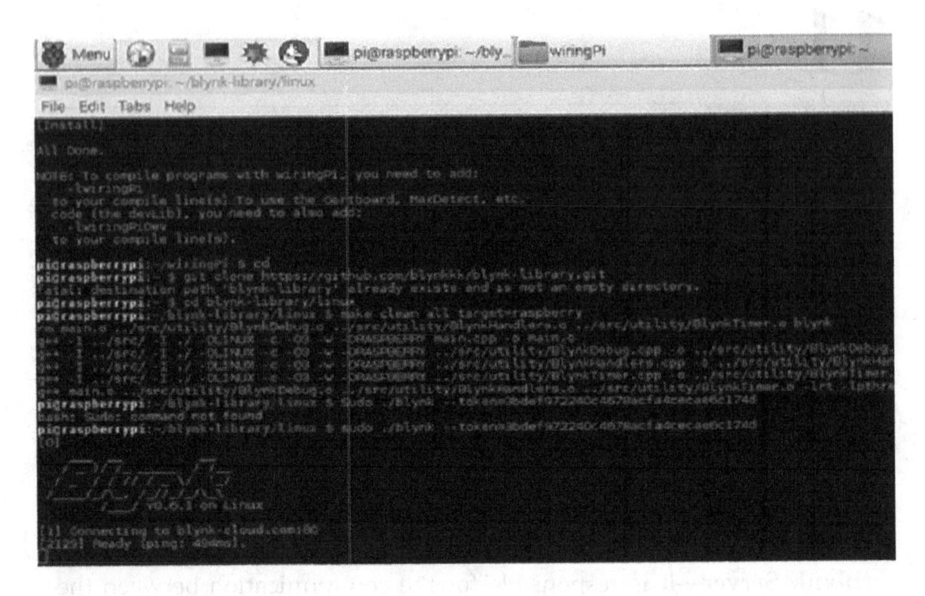

FIGURE 11.1
Raspberry Pi window.

11.2 Creating New Project with Blynk

Blynk application is available for Android and IOS both.

Steps to create Blynk project

1. Download Blynk for Android or IOS, as per the requirement, and create a new account (Figure 11.2).
2. Create a new project by providing a name, controller/processor, and internet connection to it (Figure 11.3). The background of the app can also be selected as dark or light.
3. The **Auth token** will be sent to the email address. This **Auth token** will be used for connecting Raspberry Pi to the new project widget.
4. Open the blank project and add widgets as required for the project (Figure 11.4). For a simple project of making a LED ON/OFF with button by clicking on it. Select a digital pin of Raspberry Pi to which LED is connected. Here GPIO21 is used to connect LED. Figure 11.5 shows the button for LED.
5. Add one more LED button connected to GPIO20. Now two LEDs can be controlled through the designed app (Figure 11.6).
6. Connect Raspberry Pi to the Blynk widget by running the **Auth token** command and check the workings.

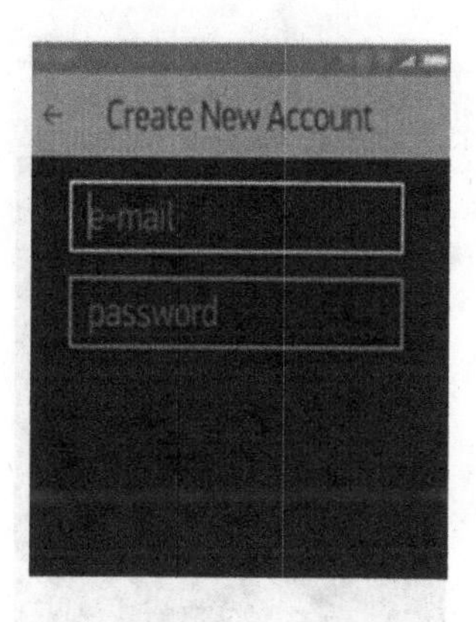

FIGURE 11.2
Creating a new account.

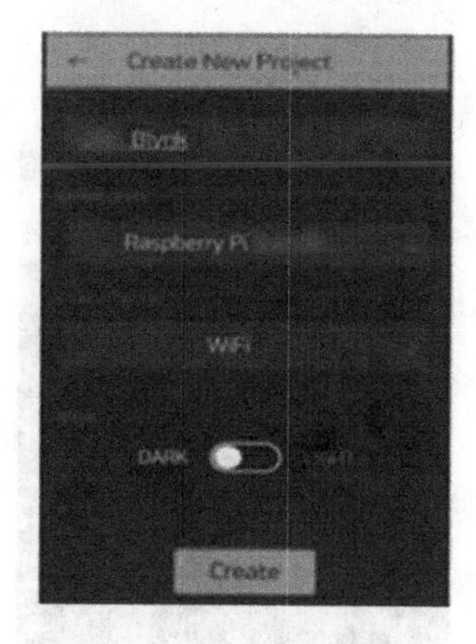

FIGURE 11.3
Create a new project.

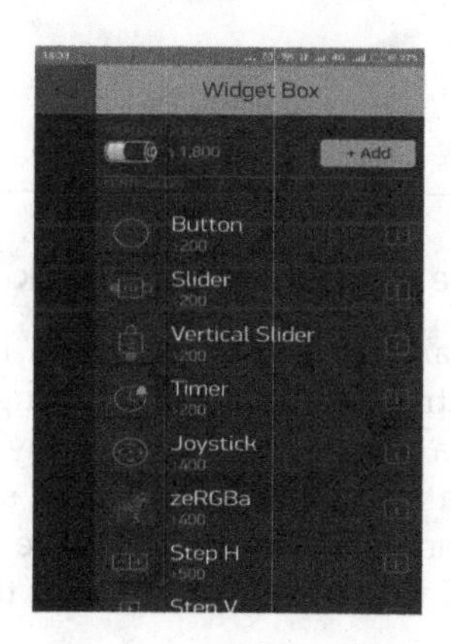

FIGURE 11.4
Widget box.

FIGURE 11.5
Button for LED.

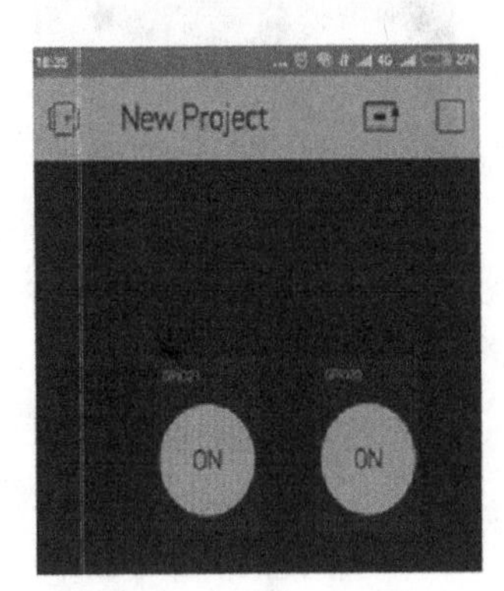

FIGURE 11.6
Project to control two LEDs.

11.3 Home Appliance Control with Blynk App

The Blynk application can be used to create an interface to control home appliances for smart control over Internet of Things (IoT). The system comprises of a Raspberry Pi, a power supply, two relay boards (if two appliances needs to be controlled), a transistor 2N2222 (as a switch), and a power supply. Connect the components as shown in Figure 11.7 and follow the steps mentioned in Section 11.2 to control two LEDs, as the same project will work to control two relays.

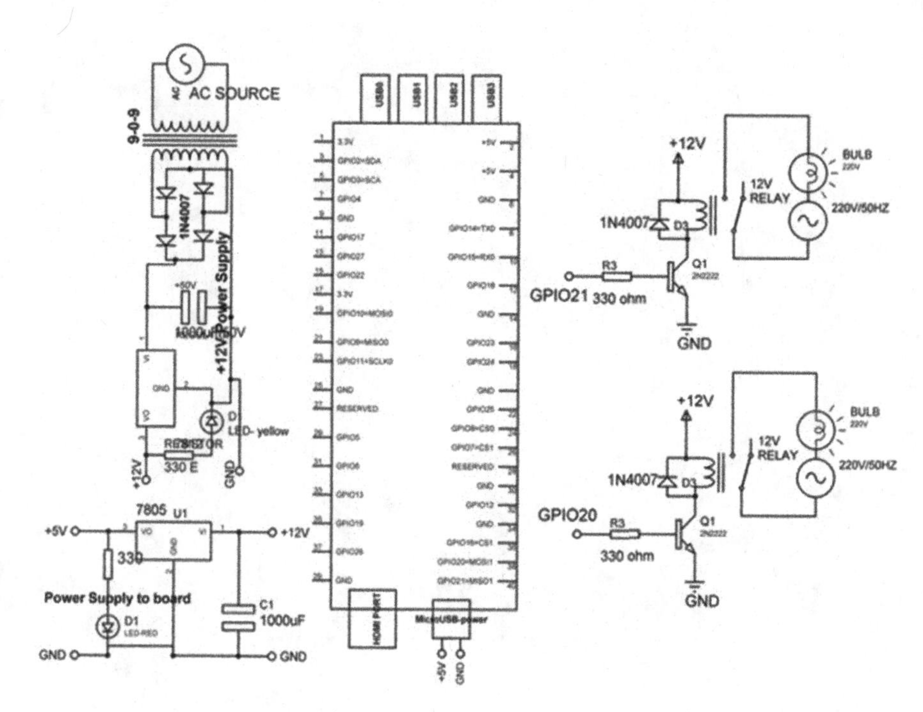

FIGURE 11.7
Circuit diagram for home appliance control.

Connections:

- Connect the base of the first transistor 2N2222 to GPIO21 through a resistor and emitter to the ground.
- Connect the base of the second transistor 2N2222 to GPIO20 through a resistor and emitter to the ground.
- Connect the collector of transistor to "L2" of relay.
- Connect the positive terminal of +12V battery to "L1" of relay.
- Connect a diode 1N4007 across "L1" and "L2."
- Connect one terminal of AC to "common" of the relay and the other to one of the terminal of AC to the load (bulb).
- Connect the other terminal of the AC load to the "NO" terminal of the relay.

12

Cayenne Application with Raspberry Pi

12.1 Introduction to Cayenne

Cayenne has a customizable dashboard with drag-and-drop widgets. It is easy to set up Cayenne, as it connects with Pi easily. In Cayenne, sensors, motors, actuators, GPIO boards, and more can be added. It is used to control the devices remotely. It has four key components: Cayenne-agent software (responsible for communicating to the server), cloud (processes and stores the sensory data), online dashboard (provides a graphical environment), and Cayenne app for Android and IOS.

12.1.1 Getting Started with Cayenne

1. Open myDevices on the Cayenne website and create an account by clicking on "Get Started for Free" (Figure 12.1).
2. Fill out the registration form and choose a device to work on (Figure 12.2).
3. Download and install myDevices Cayenne on Pi (Figure 12.3), by using commands:

   wget https://cayenne.mydevices.com/dl/rpi_8bf9u0m7hl.sh

   sudo bash rpi_8bf9u0m7hl.sh -v
4. Go to the GPIO menu and select as per the requirement of the project (Figure 12.4).

12.2 LED Blynk with the Cayenne App

The simplest way to understand the working with Cayenne app is with LED. The system is comprised of a Raspberry Pi, a power supply, and an LED. Connect the LED to GPIO17 (Figure 12.5). Follow the steps mentioned in Section 12.1.1 to control. Develop a Cayenne App. Go to the GPIO 17 pin, click on "input," and change it to "output." The status of the LED can be changed by changing the state of the button to "HIGH" and "LOW."

FIGURE 12.1
Create an account.

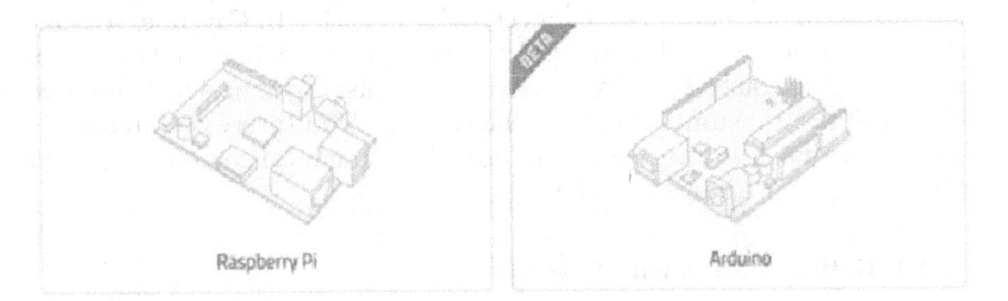

FIGURE 12.2
Choose the device.

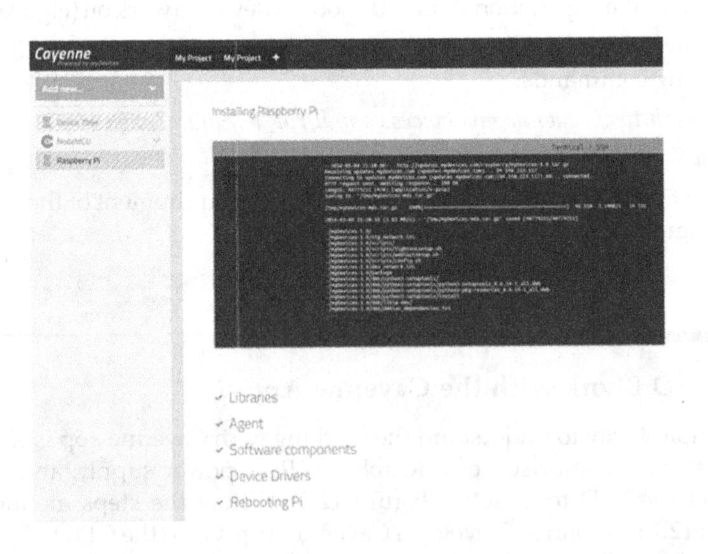

FIGURE 12.3
Installing Raspberry Pi.

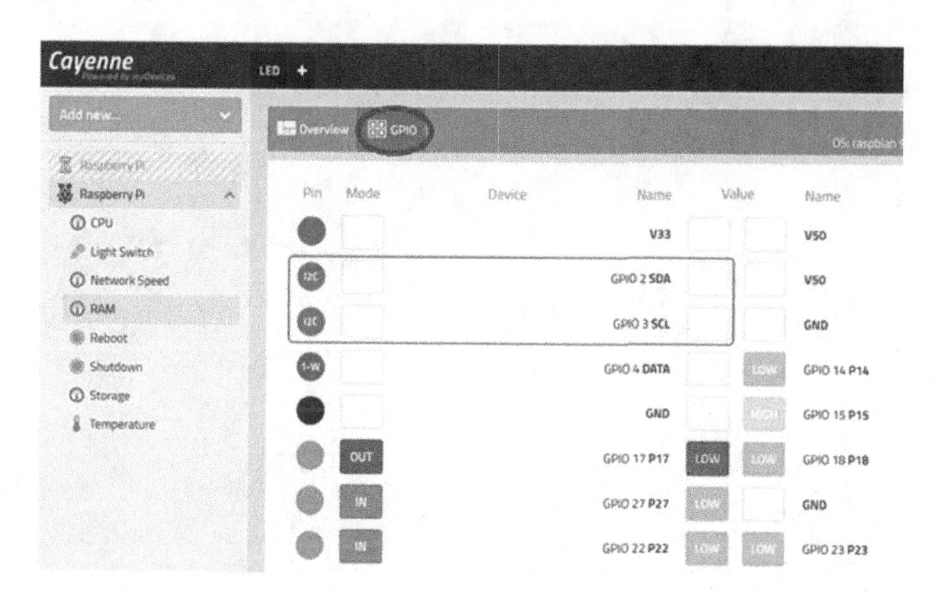

FIGURE 12.4
GPIO of Raspberry pi.

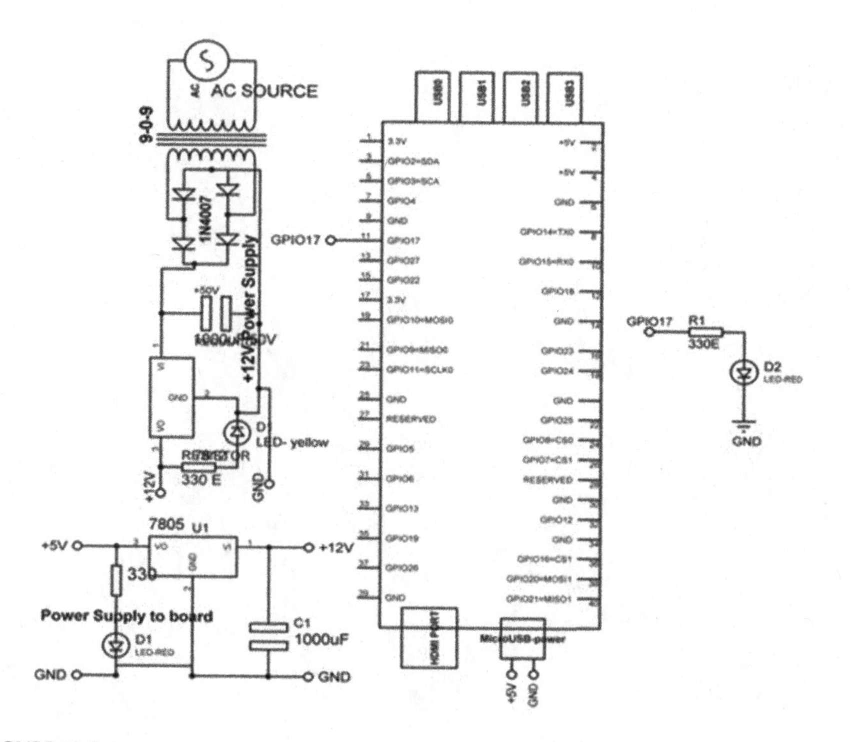

FIGURE 12.5
Circuit diagram for interfacing LED with Raspberry Pi.

Bibliography

Anita Gehlot, Rajesh Singh, Mamta Mittal, Bhupendra Singh, Ravindra Sharma, Vikas Garg, *Hands on Approach on Applications of Internet of Things in Various Domain of Engineering*. New Delhi, India: International Research Publication House, 2018.

Anita Gehlot, Rajesh Singh, Rohit Samkaria, Sushabhan Choudhury, Aisik De, Kamlesh, Air quality and water quality monitoring using XBee and internet of things, *International Journal of Engineering and Technology (UAE)*, 7.2 (2018).

Jan I. Freijer and Henk J.Th. Bloemen, Modeling relationships between indoor and outdoor air quality, Taylor & Francis Group. *Journal of the Air & Waste Management Association*, 50(2), 292–300 (2011).

Neeraj Kumar Singh, Amardeep Singh, Rajesh Singh, Anita Gehlot, Design and development of air quality management devices with sensors and web of things, *International Journal of Engineering and Technology (UAE)*, 7.2 (2018). doi:10.14419/ijet.v7i2.6.10077.

Rajesh Singh, Anita Gehlot, Bhupendra Singh, Sushabhan Choudhury, *Arduino-Based Embedded Systems: Interfacing, Simulation, and LabVIEW GUI*. Boca Raton, FL: CRC Press/Taylor & Francis Group, 2017.

Rajesh Singh, Anita Gehlot, Bhupendra Singh, Sushabhan Choudhury, *Arduino meets MATLAB…. Interfacing, Programs and Simulink*. Sharjah: Bentham Science, 2018.

Rajesh Singh, Anita Gehlot, Lovi Raj Gupta, Bhunpendra Singh, and Priyanka Tyagi, *Getting Started for Internet of Things with Launch Pad and ESP8266*. India: River Publishers, 2019.

Rajesh Singh, Anita Gehlot, Raghuveer Chimata, Bhupendra Singh, P.S. Ranjith, *Internet of Things in Automotive Industries and Road Safety*. India: River Publishers, 2018.

Rajesh Singh, Rohit Samkaria, Anita Gehlot, Neeraj Kumar, Kausal Rawat, Aisik De, Adil Rehman, Algorithm to read various sensors to detect the hazardous parameters in industry, *International Journal of Engineering and Technology (UAE)*, 7.2 (2018). doi:10.14419/ijet.v7i2.6.10060.

Index

Note: Page numbers in italic and bold refer to figures and tables, respectively.

A

Adafruit, 76
analog sensor, 104–106
architecture, 5
Arduino, 29–43
Arduino IDE, 33–37
Arduino Mega, 30, *31*, **31**
Arduino Nano, 31, *32*, **32**, **33**
Arduino Uno, 29, *30*, **30**

B

Blynk, 177–181

C

Cayenne, 183–185
challenges, 20
characteristics, 3
connectivity, 3
criteria, sensor selection for, 17
CSV file, 146–147

D

data acquisition, 145–154
DC motor, *72*, 72–74
design methodology, 19
design principles, 4
DHT11, 83–86, *84*
digital input, 66–67
digital sensor, 100–103
disaster management system, 169–173
dynamic nature, 3

F

face recognition, 60

G

gateway, 5
generation, sensors, 17

H

heterogeneity, 4
home appliance control, 180–181, *181*

I

intelligence, 3
Internet of Things (IoT), 3
IoT levels, 10–11
IPv6, 7
IR sensor, 15

L

light-emitting diode (LED), 38–40
liquid crystal display (LCD), 40–43

M

machine learning, 6
Matplotlib, 149–150
M2M, 11

N

NOOBS, 48, *48*, *49*

O

OSI (Open Systems Interconnection)
model, 6, *6*

P

Pi camera, 59
principles design, 4
protocols, 6–9
proximity sensor, 16
Pyfirmata, 117–127
Python, 115–127

R

Raspberry Pi, 45–96
Raspbian, 49–51
relay, 70–71

S

security, 4
sensors, 13–17
serial communication commands, 38
servers, 21–26
servo motor, 110–113
6LoWPAN, 7
Static IP, 52–57

T

temperature sensor, 15
terminal commands, 51–52
ThingSpeak Server, 157–168
Tkinter, 129–143
touch sensor, 16

U

ultrasonic sensor, 16
UV sensor, 16

Printed in the United States
by Baker & Taylor Publisher Services